The Power of Games

Games have long played a central role in society – actually a central role in the animal kingdom. Their play provides primary behavioral mechanisms that enable animals to learn and socialize. Indeed, "play" is a core animal activity. The principal focus of this book is on how games foster human playing, learning, and competing, including how we can design games to do this better. The author provides a wealth of real-world examples of how he created games for clients in the domains of education, energy, healthcare, national security, and transportation. He has focused on training and aiding for strategic thinking, product planning, technology development, and business operations.

The technologies underlying these games became increasingly sophisticated. This has taken on greater significance as the gaming industry has grown and prospered. Gaming revenues now dwarf film and theater. New games released gain millions of sales within a few days of release. What makes games so appealing? What is the psychology of gaming? Does it vary for card games, board games, simulation games, and online games? What makes a game successful over years? What about sports games? What sociological roles do they play in our society? Why do they claim such energy and devotion? Why are sports stars able to earn enormous contracts? What is the business of these games? Why is it expected to be increasingly lucrative? What strategies might succeed or fail? Who might be the losers and winners?

This book addresses all of these questions as well as an overarching question for society – Can online games fundamentally enhance the education of employees and students? The author is convinced they can. This requires, however, that games be designed to achieve these ends. This book is intended to contribute to understanding how to create and evaluate such games. Essentially, games enable employees and managers to play, learn, compete, and achieve in terms of knowledge and skills gained, competencies attained, customers attracted, and economic outcomes. This book explains, illustrates, and motivates investments in these pursuits to these ends.

The Power of Games

Business Impacts and Innovation Opportunities

William B. Rouse

Foreword by Rebecca Rouse

First published 2025
by Routledge
605 Third Avenue, New York, NY 10158

and by Routledge
4 Park Square, Milton Park, Abingdon, Oxon, OX14 4RN

Routledge is an imprint of the Taylor & Francis Group, an informa business

ISBN: 978-1-032-79429-7 (hbk)
ISBN: 978-1-032-79428-0 (pbk)
ISBN: 978-1-003-49192-7 (ebk)

DOI: 10.4324/9781003491927

Typeset in Garamond
by Deanta Global Publishing Services, Chennai, India

Contents

Foreword

"I have always been a gamer." So begins many a conference presentation in the discipline of Game Studies. What follows is often a slide with images from a childhood favorite, perhaps a Pokémon title, or Nintendo game. There is something about games, even within the academic setting, that invites the personal.

How refreshing that the author here – who is my own Dad – offers such a broadly ecumenical viewpoint on games, pushing past the more common and narrow definition of games today as computer-based, to instead include all manner of card and board games and even sports. How refreshing too, to be reminded that we have *all* always been gamers, with the discussion of play as a phenomenon present in this volume across not only human cultures but in the animal kingdom as well.

It is also helpful to be reminded that across these many forms of games and play, technology has often been involved, even centuries before the computer. We can think of the many luxury amusement devices such as complex automata (for example, the mechanical theater from the 1700s at Schloss Hellbrunn in Austria), the introduction of precision timekeeping as a transformative moment in sport (commented on by philosopher and cultural theorist Walter Benjamin in his observations of the 1936 Olympic Games), as well as more direct antecedents to the video game arcade machine and home console, such as Victorian slot machines like mutoscopes and nickelodeons, as well as Victorian board games and philosophical devices like stereoscopes and zoetropes, and turn-of-the-century carnival shooting galleries.

Notably, *The Power of Games* does not come from within the games discipline itself, but rather from a systems engineering perspective. This outside approach to games is part of a strong tradition. Many other disciplines have found games to be a useful playground, from cognitive science's use of games to understand human decision making, to sociology and anthropology, to uses of games as technology demos, and even experimentation with games and game structures in literature and art within movements such as Futurism and technological advances like hypertext. Games have long been used in service of other goals, with far-reaching impacts beyond the goal of play itself.

Despite these many instrumentalized, interdisciplinary engagements with games, the nature of play (let alone fun) remains difficult to pin down. Play theorist Brian Sutton-Smith has discussed this at length in his classic 1997 book, *The Ambiguity of Play*, reflecting that instead of working to resolve the disagreement between those who consider play useful (applied) or frivolous (free), it is precisely this quality of ambiguity that lies between these perspectives which must be valued as a central and unique aspect of play. When we reflect on the politics of play we can think about which characterization of play (as useful, as evolutionary, as cross-species, as intrinsically human, as frivolous) serves who. But the child at play in the playground with their friends remains blissfully unaware of such distinctions, thankfully.

The central contribution of *The Power of Games* regards the capacity of digital games to educate and shape decision making through interactive simulation. The author shares his experiences and valuable design take-aways developing and deploying simulation games across a range of fields including healthcare, education, and transportation. The larger context of the book brings pertinent philosophical reflections to the fore that touch on continually unsolved debates at the core of the games discipline: where is the intersection between games and fun? Fun and learning? When did learning need to "become" fun, and why have digital games been seen as the answer for this problem?

The Power of Games addresses these questions and others with a specific focus on the application of games to education and training in the latter part of the volume, following a broad social and historical contextualization of games. The author's breadth of inquiry regarding games for learning offers an interesting perspective, as he looks both at video games and board games used in academic settings, as well as other playful interventions in academic education such as STEM camps. Reporting on a meta-study approach to understanding these types of game- and play-based educational strategies, he concludes that the most successful interventions have a particular quality in common: a focus on design for student engagement.

An interesting contrast is mentioned between research carried out on games for simulation training as opposed to academically oriented games. Outcomes of games for training have been studied longitudinally more often, while this research method is less common in games for academic learning. There is an opportunity here for future research to help bring understanding to the long-term impacts of educational games in academic settings.

Moving into the discussion of serious games, the author describes the design of the *Immersion Lab*, drawing our attention to the importance of interface, particularly with respect to engaging users in dynamic representations of complexity. Simulations of a range of complex systems are discussed, including in healthcare, higher education, and transportation. Here we see games in service of seeking solutions to extreme challenges, such as the opioid crisis. The simulation game structure in these examples acts as a toy, allowing users to play with possible futures and understand a range of outcomes.

Using game technologies in combination with more flexible play structures that allow for the changing of rules, for example, means that the simulation sits between game and toy in a balance of constraint (real data) and flexibility (manipulatable conditions) that allows for the extension of the human imagination, just as the childhood toy extends the imagination of the child at play. And as the author notes, the simulated systems he designed enable "substantial mutual discovery" for those who interact with them together (page 113 . As with a childhood toy, the process of social play enabled by the play object reveals the meta-power of the toy: as a platform for engagement *between* humans.

Indeed, the ability to connect humans with one another in a meaningful way has been lauded as the great power of games by others. For example, Jane McGonigal's 2011 manifesto for games, *Reality is Broken: Why Games Make Us Better and How They Can Change the World,* reflects this utopic viewpoint on games. McGonigal sees games as powerful connectors of people, and as effective tools for pulling players out of reality such that they can imagine better futures together. This perspective is not unlike older, similarly optimistic views on the early computer and internet, as presented by Ted Nelson in his innovative double-sided 1974 book, *Computer Lib/Dream Machines*. Nelson's view is broader than McGonigal's in some ways, focusing not only on games but the computational medium as a whole, in terms of how it could be applied as a liberatory tool to transform complex systems such as education and economics, for young and old alike.

In contrast with these bright visions of games, there are more sober valuations of the medium, highlighting the addictive nature of gaming (including gambling), the oppressive labor practices of the games industry, and great shortcomings in terms of the role of games in culture today. In particular, Anastasia Salter and Bridget Blodgett examine traumatic incidents such as GamerGate and the detrimental bullying cultures in many online games in their 2017 volume, *Toxic Geek Masculinity in Media: Sexism, Trolling, and Identity Policing.* While the distance between these utopic and dystopic understandings of games may seem too far to bridge, we can recall the ambiguity of play emphasized by Sutton-Smith. Perhaps both perspectives are correct in that they each describe realities and potentials of games, which, while seemingly unreconcilable, the magic circle of play manages to hold them both, simultaneously.

Perhaps this apparent conflict within understandings of games can function as a productive agonism, as envisioned by political theorist Chantal Mouffe, in which diversity and conflict lead to a generative pluralism. Other media theorists, such as Jay David Bolter, are less positively disposed to the post-digital pluralism in which we find ourselves today, as discussed in his 2019 book, *The Digital Plenitude: The Decline of Elite Culture and the Rise of New Media.* In this case, the leveling of cultural production towards a more democratic field is emphasized as positive, but the seemingly endless variety of cultural and media choices today is seen as more of a mirage, with an underlying hegemony that is difficult to find a way out of, and carries with it another type of oppression.

Nevertheless, in spite of these challenges (or is it perhaps *because* of the necessary friction they provide?) liberatory practices in games persist, not least in the designs of serious games for education, training, and simulation. These threads have always been with us in games – from the early days of creative modding and self-assembly kits, to more the more recent practices in library sharing and the right-to-repair movement. Games are everywhere, and not likely to disappear. Understanding the positive potentials of the medium, across all manner of interfaces, digital and physical, is key to developing a better future for games, together. *The Power of Games* speaks to this future in a lively and informative manner, for the seriously curious reader who wishes to investigate what games can do across a broad range of contexts.

Rebecca Rouse
University of Skövde, Sweden

Preface

My family members were card players. Playing cards easily trumped television. These were serious endeavors. Winning mattered. Board games often involved children and were not taken as seriously. Sports games, particularly when it was Red Sox versus Yankees or Celtics versus Lakers, captured much attention but, frankly, not as much as four people sitting around a table, favorite drink in hand, in mortal combat.

I became fascinated with games. I still play six online games every morning – Bridge, Cribbage, Hearts, Rummy, Solitaire, and Spades – until I win every one. It typically takes 30 to 60 minutes. The cognitive dissonance between Bridge and Cribbage, for example, gives me pause as I am forced to change from one set of rules and objectives to another, which keeps me, I think, cognitively facile and adaptable, although I really don't know that.

Games have long played a central role in society, actually a central role in the animal kingdom. Their play provides primary behavioral mechanisms that enable animals to learn and socialize. Indeed, "play" is a core animal activity. The principal focus of this book is on how games foster human playing, learning, and competing, including how we can design games to do this better.

I provide a wealth of real-world examples in this book of how I created games for clients in the domains of education, energy, healthcare, national security, and transportation. My businesses focused on training and aiding for strategic thinking, product planning, technology development, and business operations. The technologies underlying these games became increasingly sophisticated.

This has taken on greater significance as the gaming industry has grown and prospered. Gaming revenues now dwarf film and theater. New games released gain millions of sales within a few days of release. What makes games so appealing? What is the psychology of gaming? Does it vary for card games, board games, simulation games and online games? What makes a game successful over years?

What about sports games? What sociological roles do they play in our society? Why do they claim such energy and devotion? Why are sports stars able to earn enormous contracts? What is the business of these games? Why is it expected to be increasingly lucrative? What strategies might succeed or fail? Who might be the losers and winners?

The Power of Games addresses all of these questions as well as an overarching question for society. Can online games fundamentally enhance education of students, employees and citizens? I am convinced they can. This requires, however, that games be designed to achieve these ends. This book is intended to contribute to understanding how to create and evaluate such games.

William B. Rouse
Washington, DC

About the Author

William B. Rouse is a Research Professor in the McCourt School of Public Policy at Georgetown University and Professor Emeritus and former Chair of the School of Industrial and Systems Engineering at the Georgia Institute of Technology. His research focuses on mathematical and computational modeling for policy design and analysis in complex public–private ecosystems, with particular emphasis on healthcare, education, energy, transportation, and national security. He is a member of the US National Academy of Engineering and Fellow of IEEE, INCOSE, INFORMS, and HFES. His recent books include *From Human-Centered Design to Human-Centered Society* (Routledge, 2024), *Beyond Quick Fixes* (Oxford, 2023), *Bigger Pictures for Innovation* (Routledge, 2023), *Transforming Public–Private Ecosystems* (Oxford, 2022), *Failure Management* (Oxford, 2021), and *Computing Possible Futures* (Oxford, 2019). Rouse lives in Washington, DC.

Chapter 1

Playing Games

Introduction

I have long been interested in computational models and simulations, and how they can be used to educate K–12 students, college students, and employees. We developed and deployed a wide range of training simulators for network management (Rouse, 2007). I discuss this series of training and research simulations in Chapter 2.

This decades-long series of studies led Ken Boff and me to consider the possibilities quite broadly. We followed a highly interdisciplinary workshop with a 700-page edited volume of contributions from a range of thought leaders, including engineers, computer scientists, and behavioral scientists. Contributors also included Hollywood producers and *The New York Times* games critic. The result was *Organizational Simulation: From Modeling and Simulation to Games and Entertainment* (Rouse & Boff, 2005).

Our ambitions blossomed. We developed and evaluated the *Health Advisor* game for pre-med students at Emory University (Basole, Bodner, & Rouse, 2013). Students interviewed patient avatars and provided them guidance on appropriate next steps – referrals – given their symptoms. We came to appreciate the work required to synthesize believable and interesting characters. Avoiding student boredom with the game was a top priority.

We decided that game players should be able to immerse themselves in the complexity of their worlds. The *Immersion Lab* resulted – see Figure 1.1. The lab was 8′ by 20′ with seven touch-sensitive screens. Decision menus and model-based projections could be displayed on any combination of the seven screens.

We applied this concept to three major studies:

- Impacts of the Affordable Care Act on the New York City health ecosystem (Yu et al., 2016)
- An economic model for exploring alternative futures of research universities (Rouse, 2016; Rouse et al., 2018)
- Impacts of Battery Electric Vehicles (BEVs) and Autonomous Vehicles (AVs) on the automobile and insurance industries (Liu et al., 2018, 2020)

I discuss these studies at length in Chapter 8. It is useful to reflect here on the overall reactions of executives and senior managers to these immersive games. They very much liked the idea of their

DOI: 10.4324/9781003491927-1

Figure 1.1 ***Immersion lab.***

teams being able to "fly the future before writing the check." They also highly valued the ways in which these highly interactive environments enabled them to "get rid of bad ideas quickly." I discuss in the following chapters the psychology, sociology, and politics of how pursuit of these objectives is facilitated.

Role of Play

The book is about "play" and its role in learning, competing, and achieving in the games of life, both for humans and other animals. This book is also about designing and orchestrating play, often termed education and training. The learning outcomes sought range from physical coordination and basic social skills to occupational competencies.

Play often commences from the moment of birth (Dugatkin & Rodrigues, 2008). Foals, newborn horses, often engage in play immediately. Needs for learning are immediate. Human children truly start exploring the world around them with sights, smells, touch, and sounds within six months of birth.

This delay, compared to horses and other mammals, reflects the reality that humans are born with brains that are largely immature, leaving babies with little control over their movements. This is the result of a lengthy evolutionary battle between big brains and narrow pelvises.

What is play? According to Tracy Treasure (2018),

> Play is a very broad term for a variety of activities and experiences that can be observed in humans of all ages, yet understandings of play and beliefs about play vary enormously. Play has long been valued in early childhood education and care—defined as contexts catering for children from birth to eight years of age—and the importance of play to young children's healthy development and learning is well documented and

> well researched. Play research covers a vast domain. Philosophers, theorists, psychologists and educators have been researching the topic of play and its value for centuries. But what is play? There is no simple definition of play and the borderlines around play, work, and academic learning are not always clear and vary according to personal beliefs. Play can be viewed as the natural vehicle by which young children learn (Wood, 2007), yet may be pushed aside in favor of work or more formal academic learning (Kernan, 2007).
>
> Over the years play has been interpreted as many things. Play has many definitions, characteristics, approaches, categories and types. It is probably easier to compile a list of play activities than it is to define play—no one definition of play can encompass all the views, perceptions, experiences and expectations that are connected with it (Kernan, 2007). Play maybe somewhat difficult to define but, nevertheless, there appears to be broad agreement among theorists coming from multidisciplinary perspectives that play makes an important contribution to children's development. Regardless of how the word is lived out in action, play has always been viewed as beneficial to the learning and development of the child.

My sense is that all animals, of all ages, enjoy play. Hide and seek, and tag, are replaced by other types of play as we age, e.g., cards, board games, sports. I have friends who creatively orchestrate "joke nights" – they are Irish. These playful events are thoroughly enjoyed by all participants. These events are fun.

Value of Fun

What is fun and why do we seek it? Catherine Price provides an insightful characterization.

> True fun materializes when we experience the confluence of three psychological states: playfulness, connection and flow. Playfulness is a quality of lightheartedness that allows you to do things in everyday life just for the pleasure of it. Connection refers to the feeling of having a special, shared experience with another person. Flow describes the state of being fully engaged and focused, often to the point that you lose track of time.
>
> **(Price, 2021, 2023)**

These constructs provide help to distinguish the types of fun enabled by the games I consider in this book. All of the games I consider tend to be addressed playfully. These games, if one chooses to engage, can be pleasurable. Admittedly, they can also be frustrating. For example, my golfing colleagues have often admitted substantial frustration.

Connection can help to overcome this, as golfing buddies seem to truly enjoy being together. My experience of connection involves watching sporting events in pubs with other aficionados, even folks who are just occasional relationships rather than good friends. Talking about the play of the game, including disagreeing, can be enjoyable good fun.

Flow, a state of being fully engaged and focused, differs by context. I usually encounter this when deeply immersed in learning something and using that knowledge to perform tasks whose outcomes I highly value. This also can happen when working with colleagues to master challenges.

If asked while immersed in this process if I am having fun, I would likely dismiss the question. In retrospect, after success, I would agree.

Trying to master difficult challenges as an adult is quite different than the joy of a child playing tag or hide and seek. I think that the fun, at least for me, is gaining deep expertise and exercising it to accomplish something I care about. I experience playfulness and connection on joke nights, but much less flow.

This distinction is important to many of the cases discussed in this book. Most of these cases involve gaining new knowledge and skills, and then mastering them to achieve valued outcomes. This can range from mastering new mathematical methods to figuring out policy initiatives that can successfully address significant public challenges.

Nature of Games

There are many types of games, as outlined in Chapter 2. They differ in terms of a variety of attributes, as summarized in Figure 1.2. The nature of game play can differ substantially, in particular in terms of how games are scored. Achieving the highest score is rather different from trying to achieve the lowest score.

Game venues differ as well, at least in terms of nomenclature. Not surprisingly, the equipment needed to play games differs as one would expect. A deck of playing cards is not useful to baseball players. Hopefully, a baseball bat would not be of much use to card players.

The most important implication of Figure 1.2 is the enormous diversity of the games addressed in this book. I will not address redesigning baseball, basketball, or football, but it is important

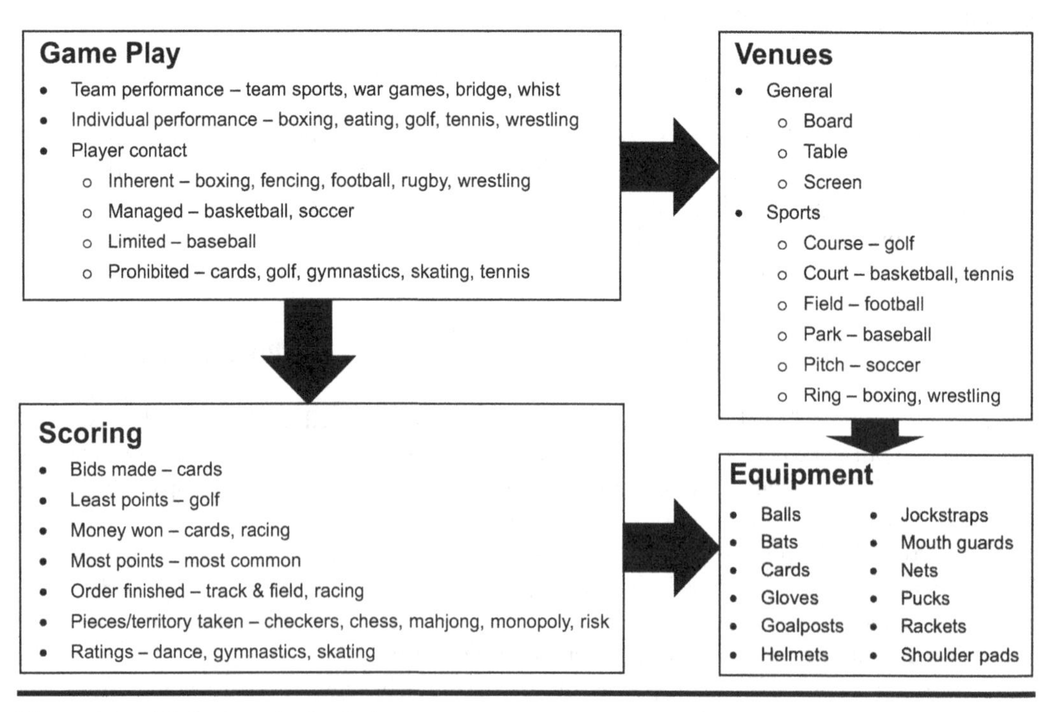

Figure 1.2 Attributes of selected games.

to keep inherent differences among games in mind. As will be clear with many of the examples I address throughout this book, it is essential to keep in mind what game you are playing.

Overall Proposition

This book is about the confluence of technology opportunities and scientific knowledge. The technology opportunities include enormous inexpensive computing power, access to extensive and rapidly growing data sets and, of course, the promise of artificial intelligence, e.g., Chat GPT. Might these technologies revolutionize education and training?

To address this question, I review the scientific evidence of what interventions have demonstrably improved education and training outcomes. After a two-year intense study of this evidence by an expert multi-disciplinary team, I have to report that the evidence is promising but far from compelling. I review these findings in Chapters 4–8. Too many studies have focused solely on students' opinions rather than educational outcomes.

Yet, I am an inherent optimist. Game technologies present burgeoning opportunities to revolutionize education and training. Chapters 9–10 characterize this opportunity space. To portend these findings, the game industry is now five times the size of the film industry. Games are developed and released first – movies are then filmed in the games.

These developments are exciting, but I am more interested in transforming K–12, college, and employee training and education. This will constitute a major contribution to fostering a healthy, educated, and productive population that is competitive in the global marketplace.

Overview of Book

To conclude this introductory chapter, I want to provide a brief preview of the intellectual journey we will take together.

Chapter 1: Playing Games

This chapter outlines my motivation for this book, succinctly summarizing the experiences that drew me to this topic, experiences that are elaborated in later chapters. Central concepts are play and fun, the nuances of which I outline. The nature of these concepts changes significantly for differing types of games, e.g., board games for children versus training programs for skilled workers. The nature of games is summarized in terms of game play, scoring, venues, and equipment. This chapter concludes with the exposition of the overall proposition that I pursue in this book.

Chapter 2: Types of Games

Fourteen well-known card games, board games, and simulation games are considered in terms of learning objectives supported by these games. Twelve training simulators are discussed, including the broad category of vehicle and process simulators, that were developed and deployed to support personnel involved with managing complex networks in a range of domains. Popular online games, sports games, and war games are characterized in terms of game play, scoring, venues. Salient differences across this range of games are discussed.

Chapter 3: History of Games

The history of games is very rich, going back many millennia. This chapter is not so ambitious to cover this wealth of developments. Instead, I consider the games discussed in Chapter 2 in terms of when and where they were developed and the profiles of their popularity. To the extent that interest in any of these games faded, I discuss how they lost to competitors. This can provide insights into strategies for sustaining interest in games, a topic of great interest to the players in Chapter 9.

Chapter 4: Psychology of Games

The psychology of games concerns human–game interactions. This includes how people understand games, develop strategies, and learn how to succeed. Motivation, engagement, and commitment to game playing are central issues for game developers. This chapter provides an extensive review of research studies that have addressed these issues. A summary of findings and their implications provides essential guidance for game designers and those evaluating games.

Chapter 5: Sociology of Games

The sociology of games considers human–human interactions in the context of game play. The first half of this chapter addresses the sociology of sports games, which play central roles in many societies. Games as social outlets are an important consideration, particularly in terms of social affiliations with sports teams. The second half of this chapter concerns the sociology of online games. Important considerations include how teams of players come together and players' relationships beyond the games.

Chapter 6: Politics of Games

Games that cross political boundaries can precipitate political controversies. Domestic rivalries such as the Red Sox versus Yankees can be vituperative, but not on the scale of controversies that cross international boundaries. This chapter addresses the political implications of three sports – Chess, the Olympics, and World Cup soccer. These sports seem to deeply engender national pride or possibly melancholy, depending not only on who wins, but also on who is judged respectful or not.

Chapter 7: Games for Learning

Games can play a central role in educating everybody. Indeed, younger people now spend more time playing games than watching television. This chapter addresses the education of three populations: K–12 students, college students, and employees. Each population has three subpopulations, yielding a total of nine targeted populations. Learning objectives for each of these populations are discussed. Alternative game-based interventions for achieving these objectives are considered.

Chapter 8: Serious Games

Serious games are designed for primary purposes other than pure entertainment. The idea shares aspects with simulation in general, but explicitly adds elements of fun and competition. This chapter reviews three serious games in detail in terms of development, deployment, and impacts. These games include applications addressing impacts of the Affordable Care Act on the New York City health ecosystem, economic impacts of alternative futures of research universities, and impacts of Battery Electric Vehicles (BEVs) and Autonomous Vehicles (AVs) on the automobile and insurance industries.

Chapter 9: Game Industry

The nature of the game industry is outlined, including its relationships with the film industry and theme parks. The most popular games engines are compared, including Unreal, Unity, Second Life, and Glass Box. Popular game stations are described, including PlayStation, Xbox, and Switch. The rapidly evolving competitive landscape is characterized in terms of market forces and technology trends. The apparent strategies of the major players are summarized.

Chapter 10: Future of Gaming

There is no doubt that gaming has a future. Its rich history is, at the very least, a strong predictor of people playing, learning, competing, and achieving. However, to what extent can games become predominant in supporting the target populations to achieve the learning objectives discussed in Chapter 7? While this possibility makes great sense, there are traditional forces that will ardently support the status quo of education and training. This chapter discusses mechanisms for mitigating such opposition and gaining the benefits that gaming promises.

References

Basole, R.C., Bodner, D.A., & Rouse, W.B. (2013). Healthcare management through organizational simulation. *Decision Support Systems*, 55, 552–563.

Dugatkin, L.A., & Rodrigues, S. (2008). Games animal play: Animal play is serious business, say scientists. *Greater Good Magazine*, March 1.

Kernan, M. (2007). *Play as a Context for Early Learning and Development: A Research Paper.* Dublin, Ireland: National Council for Curriculum and Assessment.

Liu, C., Rouse, W.B., & Belanger, D. (2020). Understanding risks and opportunities of autonomous vehicle technology adoption through systems dynamic scenario modeling – The American insurance industry. *IEEE Systems Journal*, 14 (1), 1365–1374. https://doi.org/10.1109/JSYST.2019. 2913647.

Liu, C., Rouse, W.B., & Hanawalt, E. (2018). Adoption of powertrain technologies in automobiles: A system dynamics model of technology diffusion in the American market. *IEEE Transactions on Vehicular Technology*, 67 (7), 5621–5634.

Price, C. (2021). *The Power of Fun: How to Feel Alive Again.* New York: Dial Press.

Price, C. (2023). Why we all need to have more fun. *The New York Times*, June 22.

Rouse, W.B. (2007). *People and Organizations: Explorations of Human-Centered Design.* New York: Wiley.

Rouse, W.B. (2016). *Universities as Complex Enterprises: How Academia Works, Why It Works These Ways, and Where The University Enterprise Is Headed.* New York: Wiley.

Rouse, W.B., & Boff, K.R. (Eds.). (2005). *Organizational Simulation: From Modeling and Simulation to Games and Entertainment.* New York: Wiley.

Rouse, W.B., Lombardi, J.V., & Craig, D.D. (2018). Modeling research universities: Predicting probable futures of public vs. private and large vs. small research universities. *Proceedings of the National Academy of Sciences*, 115 (50), 12582–12589.

Treasure, T. (2018). What is play? In Robinson, C., Treasure, T., O'Connor, D., Neylon, G., Harrison, C., & Wynne, S. (eds.), *Learning Through Play: Creating a Play-Based Approach within Early Childhood Contexts* (Chapter 2). Oxford, UK: Oxford University Press.

Wood, E. (2007). New directions in play: Consensus or collision? *Education*, 35(4), 309–320.

Yu, Z., Rouse, W.B., Serban, N., & Veral, E. (2016). A data-rich agent-based decision support model for hospital consolidation. *Journal of Enterprise Transformation*, 6 (3/4), 136–161.

Chapter 2

Types of Games

In this chapter, I outline over 30 popular games, all of which I have played. I consider six well-known card games, six board games, and six simulation games in terms of learning objectives supported by these games. Twelve training simulators are discussed, as well as the broad category of vehicle and process simulators, that were developed and deployed to support personnel involved with managing complex networks in a range of domains. Six popular online multi-player games and six sports games are characterized in terms of game play and origins. I provide a brief overview of war games as a precursor to more in-depth discussions in Chapters 7 and 8. I conclude with an outline of the salient differences across this range of over 40 games and simulations.

My intention in this chapter is to afford readers with a very broad view of games. Chapter 3 provides an overview of the history of these and other games. Online multi-player games are just one of seven categories considered. Games can be powerful in many ways. They can enhance one's analytic skills. They can enable understanding dynamic, uncertain phenomena. They can provide a means to enhance collaborative skills. Games can help one gain essential knowledge and practice skills central to successful job performance. We are indeed fortunate that so many types of games are available to support our pursuit of these learning objectives.

Card Games

Table 2.1 summarizes the learning objectives of six card games. I play these six online games every morning until I win each of them. It typically takes 30 to 60 minutes. The cognitive dissonance between Bridge and Cribbage, for example, gives me pause as I am forced to change from one set of rules and objectives to another, which keeps me, I think, cognitively facile and adaptable, although I really don't know that. Nevertheless, card games are in my cultural bloodstream.

I have found that playing card games with children can substantially enable the learning outcomes in Table 2.1. Counting skills are a primary benefit, especially with Cribbage. Knowing who has played what, either as a draw or discard, enhances inferential skills regarding opponents' strategies for their current hand. This also greatly helps for estimating the likelihoods of particular cards appearing by, for example, knowing that all the aces have been played. Overarching all of this is observing the pure joy children experience from the knowledge and skills they gain.

DOI: 10.4324/9781003491927-2

Table 2.1 Learning Outcomes for Selected Card Games

Games	*Learning Outcomes*
Bridge	Card probabilities, bidding, what other players hold, what cards have been played and what cards remain, logic of getting from your hand to dummy and back
Cribbage	Card probabilities, discarding to crib, maximizing your count value, minimizing your opponent's count value, counting
Hearts	Card probabilities, what other players hold, minimizing points unless you can win them all
Rummy	Card probabilities, what your opponent holds, discarding to discard pile
Solitaire	Card probabilities, maximizing card transfers from right to left stacks
Spades	Card probabilities, bidding, what other players hold, what cards have been played and what cards remain

Table 2.2 Learning Outcomes for Selected Board Games

Games	*Learning Outcomes*
Chess	Capturing the opponent's king by checking it in an unescapable position; inferring opponent's strategy to checkmate your king
Clue	Identifying the murderer, weapon, and location; keeping track of who, what, and where are no longer feasible
Monopoly	Acquiring properties, investing in houses and hotels, managing inflows and outflows of funds, forming alliances
Risk	Placing troops, planning and executing attacks, pursuing secret mission, forming alliances
Scrabble	Vocabulary, point counts, other players' likely letter holdings, options for playing difficult to play letters
Sorry	Counting, positioning multiple pieces to minimize chances of being knocked back, maximize chance to knock others back

I experienced this with my aunt and grandmother, who taught me several games. My grandmother and I played Canasta endlessly. I progressively got better. She finally refused to play with me as I always won. I had similar experiences with my aunt for other games. Beyond winning, I learned that if I applied myself, I could continually improve. I have applied that lesson to many aspects of life.

Board Games

Learning outcomes for selected board games are summarized in Table 2.2. These games often have another level of complexity compared to card games. There is a playing board and usually many more rules. In addition, all the players can see the overall state of the game. I have played most of these games with my children, except for chess and Risk that I usually played with adults.

My children were particularly attracted to Monopoly, Scrabble, and Sorry. They quickly moved beyond Sorry, but Monopoly and Scrabble continue their attraction. All-day Monopoly games still occur with some regularity. Compared to their younger years, the appetizers and beverages have substantially evolved. The intensity with which they try to gain control of Atlantic City has, if anything, grown substantially. The emotions seem to go beyond having fun.

My experiences with Risk are of particular note. I usually played the game in South Africa, where I frequently traveled on business. The players were Afrikaners, Israelis, and me. The game would take most of a day, laced with a steady stream of appetizers and drinks. People would walk outside, perhaps on the porch, to make deals for mutual conquests. Many were completely unscrupulous in honoring deals. As a result, they often won. This provided compelling lessons for the naïve American.

Beyond the types of board games in Table 2.2, this approach to gaming has broad application in learning games and serious games. Players around a table with a playing board, taking turns to move pieces in collaboration or competition provide a fairly broad gaming paradigm. Of course, much of this now happens on screens, although tables are still useful for appetizers and beverages.

Simulation Games

Once you move from boards to screens, the mechanisms for advancing states of games can draw upon a much richer set of computational means. Table 2.3 summarizes the leaning outcomes for selected simulation games, one of which I developed and deployed. The state of the game is advanced computationally rather than by moving game pieces.

Grand Theft Auto and *Need for Speed* are similar in that players are competing with authorities. *Myst* is quite different, as it took me quite a while to figure out what the game was actually about. *SimCity* provides a "god's eye" view of a city, which the player gets to develop. *The Sims*, by the same creator as *SimCity*, Will Wright, enables social interactions with no overarching goal.

Table 2.3 Learning Outcomes for Selected Simulation Games

Games	*Learning Outcomes*
Grand Theft Auto	Controlling criminals and attempting to increase their standing by completing missions in various virtual cities
Health Advisor	Interviewing patients, interpreting their expressed symptoms, and learning what services they need next
Myst	Linking books from the main hub island to reach other ages, determining who to aid in the other ages, and returning
Need for Speed	Completing street races, avoiding local law enforcement, proprioceptive responses of vehicle
SimCity	Dividing city into zones, add buildings, change tax rate, build power grid, build transportation systems, etc.
The Sims	Engaging fully in an interactive social environment, make choices, and learn how others react to these choices

All of these games are focused on having fun. The *Health Advisor* game, as mentioned in Chapter 1, was developed and evaluated to train pre-med students at Emory University (Basole, Bodner, & Rouse, 2013). Students interviewed patient avatars and provided them guidance on appropriate next steps – referrals – given their symptoms. We came to appreciate the work required to synthesize believable and interesting characters. Avoiding student boredom with the game was a top priority.

Training Simulators

Most people are familiar with the concept of aircraft simulators, where pilots can learn about displays and controls, flight management systems, normal procedures, emergency procedures, and proprioceptive responses of aircraft. There are also automobile simulators that can provide similar training. I have conducted research in nuclear power plant simulators where operators can learn about and practice procedures and a range of decision support functions. Thus, training simulators serve as learning games, a topic I return to in Chapter 7.

We developed the 12 simulators for understanding and managing the network ecosystems summarized in Table 2.4. These simulators were developed and evaluated by PhD students that I advised, including Dick Henneman, Russ Hunt, Bill Johnson, Chen Liu, Dave Morehead, Nancy Morris, and Annie Yu, as well as colleagues Phil Duncan and Paul Frey.

The first nine of these simulators are summarized in Rouse (2007). The last three are discussed in detail in Chapter 8 and reported in Yu and colleagues (2016) and Liu and colleagues (2018, 2020), respectively. All of them helped users to learn about and manage networked relationships in these ecosystems. These simulators also enabled the development and evaluation of decision support concepts to enhance network management performance.

It is important to note that all these ecosystems were actually networked environments, with well-defined nodes and connections. Thus, the simulations were not just convenient abstractions. They represented the actual physical and computational environments of these ecosystems. This enabled a much easier transfer of training from the simulators to the actual environments. This is very important because the simulators were not intended to just result in improved performance in the simulators.

This objective is rather different than for games whose central purpose is fun. While fun motivates and enables player engagement, this is only a necessary but not sufficient outcome for training simulators. Learning in the simulator needs to lead to achievement in the real world.

Online Multi-Player Games

There are roughly 3 billion people, globally, who play multi-player online games. The global gaming market size was valued at US$250 billion in 2022 and is projected to grow to US$666 billion by 2030, yielding a compound annual growth rate of over 13%. In contrast, the movie and video production industry is about one tenth of this size, and growing much more slowly.

The global automobile market is almost US$3 trillion. The global life and health insurance industry is US$4.6 trillion, the largest global industry. Thus, the games industry has a way to go. With Microsoft acquiring Activision and Amazon's entry with Amazon Studios, the major players are betting on continued strong growth.

Table 2.4 Training Simulators for Network Ecosystems

Name	*Context*	*Target Population*	*Learning Objectives*	*Measured Outcomes*
TASK	Context-free networks	Commercial airframe & powerplant maintenance trainees	Increase speed & accuracy of correct failure diagnoses	Successful transfer of knowledge & skills to real equipment
FAULT	Context-rich networks	Army Signal Corp electronics maintenance trainees	Increase speed & accuracy of correct failure diagnoses	Successful transfer of knowledge & skills to real equipment
BGLS	Functional networks	Maintenance trainees for Navy SH-3 helicopter blade-fold mechanism	Ability to trace wires and connections in cramped underbelly of helicopter	Enhanced speed & accuracy compared to use of large paper diagrams
PLANT	Production networks	Operators of production networks.	Ability to optimize production as well as detect and diagnose equipment failures	Well-designed procedures assure superior performance
MABEL	Context-free hierarchical networks	Managers of communications and service networks	Abilities to monitor, access, browse, and evaluate network operations	Cluster size improves and number of levels degrades performance
CAIN	Context-rich hierarchical networks	Managers of communications and service networks	Abilities to monitor the system looking for evidence of failed components.	Number of levels and degrees of connectivity impact performance
TMT	Networks of team members	Anti-air team in Combat Information Center of Navy Aegis Cruiser	Increased understanding of relationships among team members' tasks	Successful transfer of knowledge & skills to full-scope Aegis simulator
DBASE	Knowledge networks	Researchers seeking evidence to support intervention decisions	Abilities to identify, trace, and interpret evidence streams	Decision support to assess evidence streams enables improved performance
ILLINET	Public service networks	Managers of information services networks	Understanding of how best to managing routing and servicing of requests	Simulation of network routing protocols enables optimizing costs

(*Continued*)

Table 2.4 (Continued) Training Simulators for Network Ecosystems

Name	*Context*	*Target Population*	*Learning Objectives*	*Measured Outcomes*
NYC	Healthcare delivery networks	Executives and senior managers of healthcare provider corporations	Abilities to identify attractive merger and acquisition opportunities	Simulations enables identifying attractive and feasible candidates
BEVs	Automobile industry	Executives and senior managers of automotive companies	Understanding factors likely to most affect market growth of battery electric vehicles	Federal and state subsidies can grow market but not sustain it
AVs	Automobile industry	Executives and senior managers of automotive companies	Understanding factors likely to most affect market decline of auto insurance premiums	Decreasing accidents for autonomous vehicles will eventually cut revenues

Another interesting contrast is with the profitability of Apple, Microsoft, and Alphabet (Google). With US$237 billion in earnings and over US$5 trillion in market capitalization, they are the top three enterprises globally. Thus, the big players have both resources and intentions.

Table 2.5 summarizes the game play for selected online multi-player games that are strong contributors to the games industry revenues. *Call of Duty*, *League of Legends*, and *World of Warcraft* all center on war and battles. *Fortnite*, *Minecraft*, and *Zelda* are more survival oriented, with conflict but not formal battles. Teamwork is central to all six of these examples. Thus, there are strong social components that players value, which I address in more depth in Chapter 5.

Sports

Organized sports have long existed, certainly as far back as the original Olympic games in 776 BC. Table 2.6 summarizes the origins of selected sports games, those with which I am most familiar. These are "organized" sports, with agreed-upon rules and formal governing associations.

Business of Sports

Professional and collegiate sports are large businesses. Professional sports generate US$0.5 trillion in annual revenues. Collegiate sports generate US$16 billion annually. While many see sports as great fun, others see money to be made, particularly for teams that are privately owned. Sports are businesses and athletes are employees, which can create tensions as outlined below.

Table 2.7 summarizes how these six professional sports are governed, as well as how collegiate sports are managed. Professional and collegiate sports are interwoven because colleges and universities, in effect, provide the farm teams for professional sports. Baseball and hockey have minor league systems, but the vast majority of players drafted by professional teams come from collegiate sports.

Table 2.5 Game Play for Selected Online Multi-Player Games

Name	*Released*	*Game Play*
Call of Duty	2003	Originally focused on the World War II setting, transition to a modern setting, and proved to be the breakthrough title for the series.
Fortnite	2017	After a fluke storm appears across Earth, causing 98% of the population to disappear, the survivors are attacked by zombie-like "husks" and must command home base shelters, collecting resources, saving survivors, and defending equipment.
League of Legends	2009	Two teams of five players battle in player-versus-player combat, each team occupying and defending its half of the map.
Minecraft	2011	Players explore a blocky, procedurally generated, three-dimensional world with virtually infinite terrain and discover and extract raw materials, craft tools and items, and build structures, earthworks, and machines, while fighting hostile mobs, as well as cooperating with or competing against other players in the same world.
World of Warcraft	2004	Players create a character avatar and explore an open game world in third- or first-person view, exploring the landscape, fighting various monsters, completing quests, and interacting with non-player characters or other players.
Zelda	1986	A courageous young man of the elf-like Hylian race, and Princess Zelda, a magical princess who is the mortal reincarnation of the goddess Hylia, fight to save the magical land of Hyrule from Ganon, an evil warlord turned demon king, who is the principal antagonist of the series.

Controversies

Profits from Sports. Collegiate sports, particularly men's basketball and football, generate huge revenues and, if one listens to university presidents, enormous goodwill among alumni. It is also argued that the profitability of these sports covers the costs of Title IX commitments to women's athletics.

There are much data on this topic and little of it supports these assertions. Desrochers (2013) reviews these data and reaches the following conclusions:

- In 2010, median athletic spending was nearly US$92,000 per athlete; academic spending per full-time equivalent student was less than US$14,000 for Division IA universities
- Most Division I athletic programs rely on subsidies from their institutions and students. The largest per-athlete subsidies are in those subdivisions with the lowest spending per athlete.
- Athletic costs increased at least twice as fast as academic spending on a per capita basis across each of the three Division 1 subdivisions
- There is little to mixed evidence to support assertions that winning athletic teams lead to better student applicant pools, greater alumni giving (for other than athletics), or regional economic boosts

Table 2.6 Origins of Selected Sports Games

Sport	*Origins*
Baseball	A group of New York City men founded the New York Knickerbocker Baseball Club in 1845. One of them – volunteer firefighter and bank clerk Alexander Joy Cartwright – codified a new set of rules that would form the basis for modern baseball, calling for a diamond-shaped infield, foul lines and the three-strike rule. He also abolished the dangerous practice of tagging runners by throwing balls at them. The first officially recorded baseball game took place on June 19, 1846, in Hoboken, New Jersey.
Basketball	Created by Dr James Naismith in 1891 in Springfield, Massachusetts, to condition young athletes during cold months. The objective of the game was to throw the basketball into fruit baskets nailed to the lower railing of the gym balcony. Every time a point was scored, a ladder could be brought out to retrieve the ball. After a while, the bottoms of the fruit baskets were removed. The first public basketball game was played in Springfield, Massachusetts, on March 11, 1892.
Football	American football can be traced to early versions of rugby football and association football, which have their origin in multiple varieties of football played in the United Kingdom in the mid-19th century, in which a football is kicked at a goal or kicked over a line, which in turn were based on the varieties of English public school football games descending from medieval ball games. Rule changes were instituted by Walter Camp, a Yale University athlete and coach who is considered to be the "Father of American Football."
Golf	Modern golf developed in Scotland from the Middle Ages onward, gaining popularity in the late 19th century, when it spread into the rest of the United Kingdom and then to the British Empire and the United States. A golf-like game is recorded as taking place on February 26, 1297 where the Dutch played a game with a stick and leather ball. The winner was whoever hit the ball with the fewest strokes into a target several hundred yards away.
Soccer	The history of soccer dates back more than 2,000 years ago to ancient China. Greece, Rome, and parts of Central America, all claim to have started the sport. England transitioned soccer, or what the British and many other people around the world call "football," into the game we know today. The English are credited with recording the first uniform rules for the sport, including forbidding tripping opponents and touching the ball with hands.
Tennis	Tennis originated in the monastic cloisters in northern France in the 12th century. The ball was then struck with the palm of the hand. Rackets came into use in the 16th century. The game began to be called tennis and was popular in England and France. The medieval form of tennis is "real tennis." It spread in popularity throughout royalty in Europe, reaching its peak in the 16th century.

Thus, only a handful of Division I athletics programs produce a surplus. The vast majority of programs have to be subsidized by academic revenues. In addition, many of the philanthropic gifts from pleased alumni are directed to the support of athletics. If the current debate about compensating college athletes for their services results in compensation, the economics of college athletics will suffer further – at least from the university's perspective.

Table 2.7 Governance of Selected Sports Games

Association	*Acronym*	*Founding*	*Governance*
Major League Baseball	MLB	1876	Team Owners
National Basketball Association	NBA	1946	Team Owners
National Football League	NFL	1920	Team Owners
Federation International of Football Associations	FIFA	1904	Board
Professional Golf Association	PGA	1929	Board
United States Tennis Association	USTA	1881	Board
National Collegiate Athletics Association	NCAA	1906	Board

It is useful to be realistic about the economic impact of athletics. Nevertheless, athletic expenditures are not a major driver of the escalating costs of higher education. Of course, the fact that professional sports are free riders on collegiate sports suggests that universities are subsidizing professional teams.

Segregation in Sports. Professional sports were long segregated. They slowly came to welcome black athletes. Initial success eventually came in football (1920), hockey (1926), tennis (1945), baseball (1947), basketball (1950), and golf (1963). Nevertheless, there is still a dearth of black head coaches in the sports reviewed here.

Harry Edwards published *The Revolt of the Black Athlete* in 1969. It chronicles the Olympic Project for Human Rights, which facilitated the Black Power Salute protest by two African-American athletes, Tommie Smith and John Carlos at the 1968 Summer Olympics in Mexico City. This book was published again in 2017 with an extensive new introduction (Edwards, 2017). I address this in more depth in Chapter 5.

Edwards argues that collegiate sports programs operate like plantations. Taylor Branch (2011) echoed this assessment, suggesting that colonialism may be a better metaphor. Athletes' labors, predominantly black in the high-revenue sports, generate benefits for the owners of these teams that are used to make millionaires of coaches, few of whom are black.

At the same time, black players, many of whom come from economically challenged families, have difficulties paying for bus rides and movie tickets. Edwards convincingly argues that a significant portion of these resources should be used to assure students' educational success. Despite students' carefully cultivated optimism, few of them will have successful careers in professional sports.

Compensation of Athletes. The NCAA has long had a strongly-held position on not allowing compensation of collegiate athletes. "In 1916, the NCAA designated an amateur collegiate athlete as someone that played their sport purely for the enjoyment and developing their mental, physical, moral, and social skills" (Arash, 2014).

Consequently, none of the US$16 billion of collegiate revenue noted earlier directly benefits the players whose labors generated these resources. Most of these resources come from basketball and football. These teams have predominantly black players. Thus, the NCAA policies directly conflict with the positions of Edwards and Branch.

Cracks in the NCAA barricades are beginning to emerge. Beginning in 2021, the NCAA allowed students to receive compensation for uses of names, images, and likenesses (termed NIL) by advertisers and other companies. Prior to this change, NIL earned revenues, but not for the students.

Wargames

> Wargames are analytic games that simulate aspects of warfare at the tactical, operational, or strategic level. They are used to examine warfighting concepts, train and educate commanders and analysts, explore scenarios, and assess how force planning and posture choices affect campaign outcomes.
>
> **(RAND, 2023)**

A good example of such potential applications is simulation of differing strategies with drone swarms (Schmidt, 2023). The Russia-Ukraine war has led to extensive use of drones and recognition of the benefits of this pervasive, low-cost technology. Wargaming of alternative configurations and rules of engagement are certainly in the offing.

Wargames also have educational value (Bartels, 2021). "Building school curriculums around wargaming might spark innovation and inculcate the entire Joint Force with a better appreciation and understanding of trans-regional, cross-domain multidimensional combat." However, "rather than making one-size-fits-all promises about what games can achieve, games for use in the classroom may require tailoring to specific learning objectives." This observation is relevant to all the games addressed in this book.

Twenty years ago,

> the U.S. Army released *America's Army*, a video game meant as a recruitment tool. The free-to-play tactical shooter was wildly successful, reaching 20 million players. But the servers were recently shut down—and *America's Army* has surrendered to the forces of time. While it might be facing a very honorable discharge, *America's Army* was an idea that was ahead of its time in a lot of ways. And it blazed a trail for a new type of recruiting that's still being used today.
>
> *America's Army* was only supposed to be a seven-year project, but its success encouraged the Defense Department to stay with the game, with the Pentagon spending more than $3 million a year to evolve and promote it—a drop in the bucket compared to the overall $8 billion recruiting budget.
>
> How well did it work? A 2008 study from the Massachusetts Institute of Technology found that 30% of all Americans ages 16 to 24 had a more positive impression of the Army because of the game and, even more amazingly, the game had more impact on recruits than all other forms of Army advertising combined.
>
> **(Morris, 2022)**

Conclusions

The salient differences across this range of games include physical forms of cards, boards, screens, parks, courts, fields, pitches, and courses. The rules of these games vary enormously, including how scores are kept. Players range from children and adults, to highly trained professionals.

Possibilities of injuries range from bruised egos to physical injuries and possibly death. The most risky of these games tend to be played by professionals.

Games are fun, and they have been for millennia. Yet, despite the fun, games enable us to learn, as both animals in general and people in particular. My experiences of Cribbage enabling children's arithmetic skills is a great example. Training simulators enhancing the knowledge and skills of aircraft power plant mechanics, signal corps technicians, and astronauts are also compelling illustrations.

Games can enable achieving mastery of knowledge and skills, not necessarily as a primary intention but as a by-product of learning. Beyond personal mastery, this can enable, for example, providing training simulators to the marketplace. It can also enable providing people with wonderful immersive experiences (Rouse & Boff, 2005).

This leads to the business of games. How might gaming experiences cause people to subscribe to your offerings, either for themselves or for their students or employees, and provide impressive economic returns? I am a long way from addressing this question, but we will get there in Chapters 9 and 10.

References

Arash, A. (2014). Collegiate athletes: the conflict between NCAA amateurism and a student athlete right of publicity. *Willamette Law Review*, February.

Bartels, E.M. (2021). Wargames as an educational tool. *RAND Blog*, February 8. https://www.rand.org/blog/2021/02/wargames-as-an-educational-tool.html.

Basole, R.C., Bodner, D.A., & Rouse, W.B. (2013). Healthcare management through organizational simulation. *Decision Support Systems*, 55, 552–563

Branch, T. (2011). The shame of college sports. *The Atlantic*, October.

Desrochers, D.M. (2013). *Academic Spending Versus Athletic Spending: Who Wins? Washington, DC*: American Institutes for Research.

Edwards, H. (2017). *The Revolt of the Black Athlete*. Urbana, IL: University of Illinois Press.

Liu, C., Rouse, W.B., & Hanawalt, E. (2018). Adoption of powertrain technologies in automobiles: A system dynamics model of technology diffusion in the American market. *IEEE Transactions on Vehicular Technology*, 67 (7), 5621–5634.

Liu, C., Rouse, W.B., & Belanger, D. (2020). Understanding risks and opportunities of autonomous vehicle technology adoption through systems dynamic scenario modeling – The American insurance industry. *IEEE Systems Journal*, 14 (1), 1365–1374. https://doi.org/10.1109/JSYST.2019. 2913647.

Morris, C. (2022). After 20 years, the US Army is shutting down its recruitment video game: America's Army. *Fast Company*, February 11.

RAND (2023). *Wargaming*. https://www.rand.org/topics/wargaming.html. Accessed 19 July, 2023.

Rouse, W.B. (2007). *People and Organizations: Explorations of Human-Centered Design*. New York: Wiley.

Rouse, W.B., & Boff, K.R. (Eds.). (2005). *Organizational Simulation: From Modeling and Simulation to Games and Entertainment*. New York: Wiley.

Schmidt, E. (2023). The future of war has come in Ukraine: Drone swarms. *Wall Street Journal*, July 7.

Yu, Z., Rouse, W.B., Serban, N., & Veral, E. (2016). A data-rich agent-based decision support model for hospital consolidation. *Journal of Enterprise Transformation*, 6 (3/4), 136-161.

Chapter 3

History of Games

Introduction

The history of games is very rich, going back many millennia. This chapter is not so ambitious to cover this wealth of developments. Instead, I consider the games discussed in Chapter 2 in terms of where and when they were developed. This panorama clearly illustrates how technologies have strongly influenced the evolution of games.

This evolution affected how interest in many games faded, e.g., video arcade games gave way to home game consoles, personal computers, and portable digital devices. This succession provides insights into strategies for investing in game technologies, a topic of great importance to the competitors in the games industry as I discuss in Chapters 9 and 10.

I begin by considering games that non-human animals play, mainly to illustrate the breadth of the phenomenon of play. This leads to the history of games played by human animals involving balls, cards, and boards. I summarize selected online games, including multi-player games. The history of selected sports is briefly reviewed. I then consider gaming machines, including pinball machines, video games, and online games. I conclude with a discussion of gaming cities.

Games Animals Play

Playful behaviors and forms of games are not unique to humans. A wide range of animals plays games. These games powerfully nurture animals' physical performance and social relationships with their peers.

Hooper (2023) notes that,

> Animal play is difficult to study scientifically because it's dynamic, unpredictable, and related to unobservable "internal states" of the players. Nevertheless, researchers have observed play behaviors in other primates, but also in birds, elephants, octopus, rats, and even insects. As humans do, other species may acquire important developmental skills from play including cooperation and innovation.

 DOI: 10.4324/9781003491927-3

There seems to be agreement on five criteria to determine if a behavior really is play:

- Does the behavior serve no purpose?
- Is it intrinsically rewarding (i.e., does it happen even without external rewards such as food)?
- Is it different to other behaviors that are done for a specific reason (e.g., foraging for food or searching for mates)?
- Is it repeated and not just a one-off?
- Is it initiated when the individual is relaxed?

If all these conditions are met, a behavior is considered to be play.

Hooper's examples of play include monkeys, of course, but also other primates, species of crow, parrot, elephant, dolphin, octopus, and other animals with big brains. Rats absolutely love hide and seek and become strategic masters at the game. When given colored balls, bees will voluntarily roll the ball and prefer to spend time in areas where they previously had access to ball-rolling versus areas without balls.

> Evidence for calls similar to human laughter has been found in many species; apes, monkeys, dolphins, elephants, mongooses, cows, magpies, parrots, and rats all have specific play vocalizations. Rats "giggle" uncontrollably when they are tickled by human caregivers.

Zielinski (2015) discusses five surprising animals that play, suggesting that play may help creatures establish social bonds or learn new skills. He reports on alligators repeatedly sliding down embankments, a crocodile seen surfing the waves near a beach in Australia, a Nile crocodile photographed as it repeatedly threw a dead hippo into the air, and juvenile black caimans in Brazil that were seen chasing each other in circles.

Zielinski's examples include a variety of sophisticated behaviors, including Cuban crocs "courting" each other – the male would let the smaller female climb onto him and he'd give her rides around their pool. A Nile soft-shelled turtle named "Pigface," who lived at the National Zoo in Washington, DC, was more than 50 years old at the time and was given a basketball by his keepers. As the turtle swam around the enclosure, he batted the ball in front of him.

Three male cichlid fish repeatedly attacked and deflected a thermometer in their tank. Each time a fish struck the thermometer, it moved then to fall back into place. The dominant female wasp approaches the subordinate, raising her head over the subordinate's head, performing a rapid beating of the antennae, and often licking, biting, and asking for food.

Octopuses are often among animals reported as playing. Seven adult and seven sub-adult octopuses held in a lab were given objects made of Lego blocks. Nine of the cephalopods exhibited play-like behavior, pushing or pulling the objects or floating them on the surface. Another study found octopuses blowing streams of water at empty pill bottles. The action, which causes the bottles to shoot away, qualifies as play because the octopuses did it not once or twice, but 20 times over. I highly recommend Godfrey-Smith (2016) for readers interested in octopuses.

Sharpe (2011) addresses meerkats and the question "Does play-fighting teach pups crucial combat skills (the 'practice theory') or whether meerkats that played together stayed together (the 'social bonding theory')." He suggests that "play-fighting helps youngsters learn battle skills."

Mock combat (also called "rough and tumble") is central to the games of many species and in polygynous beasts (where big macho males must brawl to secure a harem), males play-fight much more than their sisters. "Long-term research on American brown bears has revealed that cubs who romp a lot are more likely to survive to independence, even after taking into account the cub's (and its mum's) condition, and the availability of food."

Sharpe reports that

> Researchers teased apart the factors that promoted brain growth and found that sensory stimulation and arousal, even together, did not increase cortical growth unless they were coupled with interactive behavior, i.e., play or training. And it was play that had the biggest impact; in fact, the more a young rat played, the more rapidly its brain grew.

Zimmer (2006) provides another view of games animals play and observes that

> Horses kick and dance. Dogs wrestle each other to the ground. Warblers toss rocks. And when scientists give octopuses Lego blocks, they seem to have a lot of fun batting them around with their tentacles. But it is a supreme challenge to determine if they are actually playing.
>
> Certain features seem to make animals prone to playfulness. If parents feed and protect their offspring, for example, young animals don't have to always be searching for food or hiding from predators. They've got free time to fill. It also helps to have a powerful brain that can produce actions that are more than just automatic reflexes. And play is encouraged by a high metabolism, which gives animals extra energy that they can burn off doing things that are not essential to their immediate survival.
>
> Play may have first emerged in animals simply as a byproduct of their physiology. But it may have evolved into an opportunity for animals to increase their odds of survival and reproduction. Natural selection may have transformed play into a vital part of development in some animals. By playing, young animals can teach themselves about their physical environment. They can also try out complicated maneuvers that could serve them well later in life.
>
> Play is also an opportunity for animals to bond. Rats, for example, like to roughhouse with one another, releasing high-frequency chirps that may be a rodent version of laughter. It has been found by humans tickling the rats, they get these chirps in response. Studies on the brains of chirping rats indicate that chirping is accompanied by a surge in chemicals that create a feeling of pleasure. Playing together, it seems, just feels good to a rat. And the good feeling causes rats to stick together. Chirping rats tend to bond with other chirping rats and to avoid less playful ones.
>
> Some yearling coyotes drift away from their groups to live alone, while others stay behind. Fewer than 20% of the stay-at-home coyotes die, while more than 55% of the drifters do. Together, coyotes may be able to find food and fight off rivals better than they can alone.
>
> Play has a crucial role in keeping coyote groups together. Young coyotes learn how to follow social rules that can prevent conflicts from escalating into all-out battle. They can bite, but not hard enough to hurt. Play also allows coyotes to tamp down

the tensions in the coyote hierarchy. During play, dominant coyotes will roll on their backs in submission to lower coyotes.

Humans are a particularly playful species, even compared to other primates. Hominids have a longer childhood than other primates, offering more opportunity for play. Play may allow young hominids to learn survival skills, such as finding food and preparing it to eat. Play also prepares children to live in an increasingly complex social world.

Hominids evolved into highly cooperative hunters and food-gatherers and spread throughout the world. Only by learning social rules could hominid bands stick together. Our ancestors could also apply their growing powers of imagination and language to play as well. Along with good old rough-housing and stone-tossing came fairy wings and train sets.

Games of Human Animals

We are, of course, primarily interested in the power of games to entertain and support human animals. How do human and animal games differ? Parlett (2011) provides an interesting comparison.

> The most significant is that only humans play formal games. Animals play instinctively and stop playing when they get tired, hurt or interrupted; humans play by design and agreement. They are conscious of the fact that they are playing and can verbalize their agreement to have come to an end or to recognize a win.
>
> Animal play is typically categorized as locomotor, object, or social, with little or no overlap between them, but even human children can mix and match behavioral elements derived from all three, making it difficult to classify human play in such terms. The function of play in animals is to practice life-sustaining skills, which is one reason why juveniles play more than adults. (Another is that juveniles are less busy earning a living.) But this can hardly be the role of the sophisticated play practiced by human children. If not, then what is it?

"Human games are creative, imaginative scenarios in a way that is not possible for the most intelligent of animals." In addition, "Critical factors to bear in mind are the development of language and the availability of leisure." Finally, "Formal games did not spring from nowhere but must have been preceded by and built upon the foundations of many millennia of informal games."

Riede and colleagues (2017) elaborate the role of parents.

> Part of *Homo sapiens*' niche is the active provisioning of children with play objects—sometimes functional miniatures of adult tools—and the encouragement of object play, such as playful knapping (shaping by breaking off pieces) with stones. Salient material culture innovation may occur or be primed in a late childhood or adolescence sweet spot when cognitive and physical abilities are sufficiently mature but before the full onset of the concerns and costs associated with reproduction.

Thus, all animals seem to play in one way or another, but humans' games are often more elaborate and over time may become formal games. The evolution of selected human games is the primary focus of the rest of this chapter.

Games Involving Balls

The oldest known ball in the world is a toy made of linen rags and string that was found in an Egyptian child's tomb dating to about 2500 BC. In highland Mesoamerica, evidence shows that ball games were played starting at least as far back as 1650 BC, based on the finding of a monumental ball court. The oldest rubber ball found in the region dates to about 1600 BC.

The Mesoamerican Mayan ball game was played, experts think, by all the cultures in the region, beginning with the Olmecs who may have invented it. The Mayan ball game goes back 3,500 years, making it the first organized game in the history of sports. Of course, the shape of balls begs throwing, bouncing, hitting, and so on.

Card Games

Table 3.1 summarizes the origins of the card games discussed in Chapter 2. I added Whist because of its relationships with Bridge. I learned Whist before engaging in Bridge. It provided useful training.

Playing cards were likely invented during the Tang dynasty in China around the ninth century AD as a result of the usage of woodblock printing technology. They reached Europe around 1360, not directly from China but from the Mameluke empire of Egypt. The history of suitmarks demonstrates a fascinating interplay between words, shapes, and concepts. The Flemish Hunting Deck in the Metropolitan Museum of Art is the oldest complete set of ordinary playing cards made in Europe from the 15th century.

A playing card is a piece of specially prepared card stock, heavy paper, thin cardboard, plastic-coated paper, cotton-paper blend, or thin plastic that is marked with distinguishing motifs. The face and back of each card usually has a finish to make handling easier. They are most commonly used for playing card games. Cards are also employed in magic tricks, card throwing, and card houses. Playing cards usually are sold together in a set as a deck of cards or pack of cards.

Cards are often collected. However, these cards are not usually playing cards. They are often referred to as trading cards. In 1868, Peck and Snyder, a sporting goods store in New York, began producing trading cards featuring baseball teams. The cards were a natural advertising vehicle for the company's baseball equipment. The Peck and Snyder cards are sometimes considered the first baseball cards. Pokémon trading cards originated in Japan in 1996 and, by 2023, over 50 billion cards had been sold.

Table 3.1 Origins of Selected Card Games

Card Game	*Origin*	*Dates*
Bridge	England	1500s
Cribbage	England	1600s
Hearts	Spain	1750
Rummy	China or Mexico	1890s
Solitaire	Scandinavia	1700s
Spades	United States	1930s
Whist	England	1700s

I collected baseball cards in the late 1950s. I would buy a package of Topps bubble gum for 5 cents, which would include one card. Living an hour outside Boston, I hoped to get a card for Ted Williams, the Red Sox slugger. I never did and I did not care for the bubble gum.

Board Games

Table 3.2 summarizes the origins of the board games discussed in Chapter 2. Classical board games are divided into four categories: race games (e.g., Pachisi), space games (e.g., Tic-Tac-Toe), chase games (e.g., Viking Game), and games of displacement (e.g., Chess). Board games have been played and evolved in most cultures and societies throughout history.

The British Museum (2021) argues that the top ten historical board games are:

1. Royal Game of Ur, Mesopotamia, 2600 BC
2. Lewis Chessmen, Scandinavia, 1100
3. Wari, West Africa
4. Senet, Egypt, 1400 BC
5. Mahjong, China, 1600
6. Game of the Goose, Italy, 1500
7. Ajax and Achilles' Game of Dice, Athens, 530 BC
8. Sugoroku, China, 700
9. Pachisi, India, 1500
10. Mehen, Egypt, 3000 BC

The Royal Game of Ur was found in the royal tombs of Ur, dating to Mesopotamia 4,600 years ago. Senet originated in Egypt in the 3000s BC. Mehen originated in Egypt in the 2000s BC. Hounds and Jackals, another ancient Egyptian board game, originated in the 1800s BC. Backgammon originated in ancient Mesopotamia about 5,000 years ago. Ashtapada, chess, Pachisi, and Chaupar originated in India. Go and Liubo originated in China. Patolli originated in Mesoamerica played by the ancient Aztecs.

Online Games

Table 3.3 summarizes the origins of the online games discussed in Chapter 2. As I noted in discussing these games, once you move from board to screens, the mechanisms for advancing states of

Table 3.2 Origins of Selected Board Games

Board Game	*Origin*	*Dates*
Chess	India	500s
Clue	England	1940s
Monopoly	United States	1900s
Risk	France	1950s
Scrabble	United States	1930s
Sorry	England	1920s

Table 3.3 Origins of Selected Online Games

Online Game	*Origin*	*Dates*
Grand Theft Auto	United States	1997
Health Advisor	United States	2013
Myst	United States	1993
Need for Speed	United States	1994
SimCity	United States	1989
The Sims	United States	2000

games can draw upon a much richer set of computational means. These games all rely on some form of computational simulation of spaces, buildings, vehicles, etc.

These games can be fun and entertaining, as well as educational. *Health Advisor* was intended to be educational, but considerable effort was devoted to making it interesting and entertaining. *SimCity* was intended to be fun, but it was interesting to learn how to please citizens. *SimCity* motivated many other developers to create serious games as discussed in Chapter 8.

Online Multi-Player Games

Table 3.4 summarizes the origins of the online multi-player games discussed in Chapter 2. Note that all of these games have current releases despite being two decades or more from original release. In these ways, these games represent continuing franchises targeting well-understood populations of game players. These games all being online makes this feasible.

The second half of Chapter 5 concerns the sociology of online games. Important considerations include how teams of players come together and players' relationships beyond the games. Their relationships with social media add another dimension to these social systems.

Sports

Table 3.5 summarizes the origins of the sports games discussed in Chapter 2. The beginnings of sport were related to military training. For example, competition was used as a means to determine whether individuals were fit and useful for service. Team sports were used to train and to prove the capability to fight in the military and also to work together as a team (military unit).

The history of sports extends back to 70,000 BC. Study of the history of sport can teach lessons about social changes and about the nature of sport itself, as sport seems involved in the development of basic human skills, including social skills. The first half of Chapter 5 addresses the sociology of sports games, which play central roles in many societies. Games as social outlets are an important consideration, particularly in terms of social affiliations with sports teams.

Gaming Machines

In this section, I discuss gaming machines and the companies that produce them. Table 3.6 summarizes the origins of selected gaming companies and offerings. There were, of course, many other

Table 3.4 Origins of Selected Online Multi-Player Games

Online Multi-Player Game	*Origin*	*Dates*
Call of Duty	United States	2022 latest release; 2003 origin
Fortnite	United States	2023 latest release; 2017 origin
League of Legends	United States	2023 latest release; 2009 origin
Minecraft	United States	2022 latest release; 2011 origin
World of Warcraft	United States	2022 latest release; 2004 origin
Zelda	Japan	2023 latest release; 1986 origin

Table 3.5 Origins of Selected Sports

Sport	*Origin*	*Dates*
Baseball	United States	1845
Basketball	United States	1891
Football	England	1800s
Golf	Scotland	1800s
Soccer	China	1000s BC
Tennis	France	1100s

Table 3.6 Origins of Selected Gaming Companies and Offerings

Company	*Founded/Launched*	*Dates*
Atari	Founded, 1972	Acquired by Hasbro, 1998
Activision	Founded, 1979	Acquired by Microsoft, 2023
Electronic Arts	Founded, 1982	Fourth in Media & Entertainment
Nintendo NES; Switch	Launched, 1985	27.7% Market Share
Maxis	Founded 1987	Acquired by Electronic Arts, 1997
Sony Playstation	Launched, 1994	45.0% Market Share
Microsoft Xbox	Launched, 2001	27.3% Market Share

competitors that were acquired or failed along the way, as in any other industry. There are also current major players in media and entertainment, such as Disney, that are increasingly moving into gaming.

Pinball Machines

By the 1930s, the first modern-looking, coin-operated pinball machine was invented. It was created by the company Automatic Industries, which named it a *Whiffle Board*. The Bally

Manufacturing Corporation was founded by Raymond Moloney in January 1932 when Bally's original parent, Lion Manufacturing, established the company to make pinball games, taking its name from its first game, *Ballyhoo*. The Chicago-based company quickly became a leading pinball maker.

Addams Family, designed by Pat Lawlor, was the best-selling pinball machine of all time with over 20,000 units built. It was manufactured by Midway and released in 1992. The game featured custom speech from Raul Julia and Anjelica Houston and explored the strange world of the Addams Family.

My father, Gayler Rouse, loved pinball. We would have lunch at local diners that had machines. If no machine was available, my father would "buy" a machine from one of the players, giving them US$10–20 for the use of their machine. He taught me how to finesse the machines without tilting them, which would end your game. You had to carefully sense how much jostling a given machine would accept without tilting.

Video Games

Table 3.7 summarizes the origins of selected video games. These games were typically accessible in an arcade or perhaps a pub. It was unusual to have such games at home before personal computers became prevalent. In fact, video arcade games gave way to home game consoles, personal computers, and portable digital devices.

Spacewar! was a space combat video game developed in 1962 by a team led by Steve Russell. It was written for the newly installed DEC PDP-1 minicomputer at the Massachusetts Institute of Technology. The earliest known video game competition took place in October 1972 at Stanford University for the game *Spacewar!*.

Atari founder Nolan Bushnell conceived the idea of a video version of *Ping-Pong* in 1972 and assigned engineer Allan Alcorn to create it. Within several months, Alcorn developed a prototype for a coin-operated arcade version so engaging that users testing it in a neighborhood bar filled its coin box to overflowing.

Game designer named Toru Iwatani created *Pac-Man* in 1980. I visited the Swedish National Defense Research Institute in 1979–80 to give a series of lectures on human–machine systems. Bengt Bergstrom was my host. When I returned to the University of Illinois in Fall 1980, Bengt visited our research group. *Pac-Man* had just come out and was available at a local pub. We ended up having most of our research discussions at that pub, taking turns on *Pac-Man*. It was a clear hit.

Table 3.7 Origins of Selected Video Games

Video Game	*Originator*	*Dates*
Pong	Atari	1972
Space Invaders	Bally	1978
Pac-Man	Namco	1980
Donkey Kong	Atari & Nintendo	1981
Mario Brothers	Atari & Nintendo	1983

Online Games

Maze War, introduced in 1973, provided the first graphic virtual world, enabling a first-person perspective view of a maze in which players roamed around shooting at each other. It was also the first networked game, in which players at different computers could visually interact in a virtual space. In 1974, the game was enhanced so that it could be played across the ARPANET, forerunner of the modern Internet.

Adventure was created in 1975 by Will Crowther was the first widely played adventure game. The game was significantly expanded in 1976 by Don Woods. *Adventure* contained many *Dungeons & Dragons* features and references, including a computer-controlled dungeon master.

The first graphical MMORPG (massively multi-player online role-playing game) was *Neverwinter Nights* by designer Don Daglow and programmer Cathryn Mataga. The game went live on AOL for PC owners in 1991 and ran through 1997. MMO games were highly influenced by *Multi-User Dungeons* (MUDs), *Dungeons & Dragons* (D&D) and earlier social games.

The University of Illinois pioneered the PLATO system, an educational computer system based on mainframe computers with graphical terminals, influencing many areas of multi-user computer systems. By 1974, there were graphical multi-player games such as *Spasim*, a space battle game which could support 32 users, and the *Talkomatic* multi-user chat system. As an aside, I joined the faculty of the University of Illinois in 1974.

Avatar was an early PLATO game written by high school students using the PLATO system at the University of Illinois. This game ran on 512 by 512 plasma panels of the PLATO system. Groups of up to 15 players could enter the dungeon simultaneously and fight monsters as a team.

The games on PLATO were graphical in nature and very advanced for their time, but were proprietary programs that were unable to spread beyond those who had access to PLATO. Textual worlds, which typically ran on Unix, VMS, or DOS, were far more accessible to the public. *Oregon Trail*, released in 1971, is a classic example of a textual world, with still pictures. The game was intended to teach children about the realities of 19th-century pioneer life on the Oregon Trail.

Gaming Cities

Gambling in Italy existed for centuries and took on many forms and dates back to the days of the Roman Empire. The predecessor of the modern game of backgammon became popular. The first gambling house, Ridotto, was opened in Venice in 1638. It was sanctioned by the government aiming to control gambling activity of the citizens.

Admission to the gambling house was free, but only rich people could afford to play because the stakes were high. The games played were lotteries and card games. Both games had a very high house edge. In 1774, Ridotto was closed, resulting in the growth of closed gambling clubs. These clubs were called "casinos," and the word casino itself is of Italian origin. Baccarat originated in Italy as did Bingo.

The Nevada state legislature realized that gambling would be profitable for local business and legalized gambling in 1931. Las Vegas, with a small but already well-established illegal gambling industry, began its growth as the gaming capital of the world. New Jersey voters legalized casino gambling in Atlantic City in 1976, and the first casino opened two years later. Atlantic City had been the home of the Miss America pageant for many decades.

Las Vegas features many of the world's most famous and most respected casino hotels and, of course, casinos. Atlantic City is a distant second in the US. Las Vegas has over 60 major casinos, whereas Atlantic City has 30. None of the Atlantic City casinos have the name recognition of those in Las Vegas.

The Mediterranean city of Monte Carlo epitomizes European glitz and glamour. More millionaires are packed into this tiny principality than almost any other location in the world, and the marina is a berth of choice superyacht owners. It was not always this way. In the mid-1880s, the area was little more than wasteland and scrub, with a few dilapidated buildings and a near-bankrupt hotel.

How did such a dramatic transformation take place in such a short space of time? The answer is legalized gambling. Monaco is an independent principality on the border of the French Riviera and the Ligurian Coast of Italy. It has been ruled by the Grimaldi family for over 800 years, and for most of that time it made about as much of a historical impact as could be expected from a state covering only half the area of Manhattan's Central Park (Braude, 2016).

Macau, a special administrative region like Hong Kong, is the only place in China where casinos are legal, and the business has grown at an astounding pace since 2001 when the government ended the four-decade gambling monopoly of the Hong Kong billionaire Stanley Ho. The world's largest gambling destination, Macao is a one-hour ferry ride from downtown Hong Kong. A former Portuguese colony, Macao is now a semiautonomous special administrative region of China.

Horowitz (2015) contrasts Las Vegas, Monaco, and Macau. He concludes that

> While it is commonplace to believe that Monaco is the profitable venue of the past, Las Vegas is the significant environment of the present and Macau is the trendsetting location of the future, this thesis argues that the Las Vegas model is in fact the most viable framework for the future of the gaming resort industry worldwide.
>
> The ability of Las Vegas to overcome historical blunders, underworld threats and economic recessions attests to its preeminence as the strongest contemporary gaming resort model. It has shown a resilience to build and adapt in diverse conditions and markets. Its strength as a gaming resort destination is uncontested and universal.

Conclusions

The history of games is very rich. Play is central to animal cultures, human and otherwise. The card and board games I have discussed remain popular, now both in their original physical forms and online. Sports have evolved to include both physical and online forms. Nevertheless, physical sports parks, courts, fields, pitches, and courses remain popular and well attended.

Yet, change happens, as exemplified by video arcade games giving way to home game consoles, personal computers, and portable digital devices. The immense proliferation of personal and portable digital devices presents both opportunities and challenges. Technology trends can be exploited to create new gaming experiences. A central challenge is the many competitors, known and unknown, that are also doing this.

Other opportunities include games for learning (Chapter 7) and serious games (Chapter 8). Games can provide a powerful means for learning about and solving many types of problems in a wide range of domains. These games can support students, employees, and people in general to learn about new things and perhaps earn credentials.

This wealth of opportunities suggests that the numbers of providers of game-like capabilities will steadily increase. Students, employees, and consumers will come to expect such capabilities. Understanding these expectations will provide insights into strategies for investing in game technologies, a topic of great importance to the competitors in the games industry as I discuss in Chapters 9 and 10.

References

Braude, M. (2016). *Making Monte Carlo: A History of Speculation and Spectacle.* New York: Simon & Schuster.

British Museum (2021). *Top 10 Historical Board Games. https://www.britishmuseum.org/blog/top-10-historical-board-games.*

Godfrey-Smith, P. (2016). *Other Minds: The Octopus, the Sea, and the Deep Origins of Consciousness.* New York: Farrar, Straus and Giroux.

Hooper, B. (2023). The many surprising ways that animals play. *Psychology Today,* March 29.

Horowitz, D. (2015). *Monaco, Las Vegas and Macau: Gaming resorts of the past, present and future.* MS Thesis, Pomona: California State Polytechnic University.

Parlett, D. (2011). Back to square one: Questing how games began. *The Incompleat Gamester.* https://www.parlettgames.uk/gamester/backto.html.

Riede, F., Johannsen, N.N., Hagberg, A., Nowell, A., & Lombard, M. (2017). The role of play objects and object play in human cognitive evolution and innovation. *Evolutionary Anthropology,* 27, 46–59.

Sharpe, L. (2011). So you think you know why animals play. *Scientific American,* May 17.

Zielinski, S. (2015). Five surprising animals that play. *Science News,* February 20.

Zimmer, C. (2006). Games animals play. *Forbes,* December 14.

Chapter 4

Psychology of Games

Introduction

The psychology of games concerns human–game interactions. This includes how people understand games, develop strategies, and learn how to succeed. Motivation, engagement, and commitment to game playing are central issues for game developers. This chapter provides an extensive review of research studies that have addressed these issues. A summary of findings and their implications provides essential guidance for game designers and those evaluating games.

Human information processing, problem solving, and decision making underlie learning a new game and playing it successfully. How does this happen? What factors affect success or failure? And, of course, why do humans choose to do this? Further, why do humans do this repeatedly? Why do I play the same six online card games every morning for years?

This chapter addresses these questions in terms of the learning objectives summarized in Tables 2.1–2.4 for 30 different games. How do humans accomplish this learning? How can learning be enhanced? Of course, the first question is why do people engage in these games? In the next section, I address the elements of engagement.

Several human phenomena are not considered in the discussion in this chapter, in part because they are primarily relevant in sports. These phenomena include visual acuity, eye–hand coordination, and strength. Admittedly, if you cannot see the game board, or reach and grasp the game pieces, your success will be rather limited, but such fundamental limitations are beyond the scope of our discourse.

Engagement

Games in general, and formal games in particular, have agreements and rules.

> We adhere to these agreements and rules faithfully, committing to and executing the game with little hesitation. We'll only do what the rules of the game allow. We will try to win. We bring our bodies, personalities, and life experiences into the game. When the game ends, we take the memories and experience of the game with us.
>
> **(Von Ehren, 2020)**

DOI: 10.4324/9781003491927-4

> Digital games take many of the powers of traditional analog games and ramp up both the rate of interaction and the complexity of the underlying systems. A digital game takes input from the player 60 times per second, resolves it with a potentially very complicated rule set, and renders a new image of the game state. This rapid feedback loop engages our proprioception, that is our sense of embodiment and physicality. Digital games are powerfully compelling as a result, but I often find that analog games are a bit more playful. When playing an analog game, the only limitations are the rules you've agreed to, you can modify and change them at will, more easily creating playful experiences.
>
> **(Von Ehren, 2020)**

Trevisan (2023) provides a recent review of research on the psychology of motivation in games. The research reviewed tends to see motivation as involving three components, ranging from flow, fun, and presence; to engagement, engrossment, and immersion; to autonomy, competence, and relatedness. The central observation is that motivation is multi-dimensional.

Ryan and Deci's highly cited classic (2000) addresses autonomy, competence, and relatedness, which they define as follows:

- Autonomy is the will of the self to have as much control over its own actions as possible
- Competence is the striving for being able, being good and recognized in every action;
- Relatedness is the urge to connect, to interact and care/being cared for by others.

They emphasize that the fulfillment of these needs does not rely on objective judgment but on personal perception. In a subsequent article, Ryan and his colleagues conclude that "Along with expanding autonomy, games have also increasingly been apt at satisfying relatedness needs by providing opportunities for engaging in online interactions and communities" (Ryan, Rigby, & Przybylski, 2006).

Acaster (2023) provides a comprehensive perspective on why people play games.

> Different games can fulfill different categories of needs but games are often particularly good at fostering belongingness needs by allowing players to interact and grow relationships with others through playing together or sharing their interest; and satisfying esteem needs by allowing players to access feelings of accomplishment when improving or succeeding within the context of the game.
>
> Games almost always produce perceptible results and have clear, predictable achievement and reward systems, which can act as a path to fulfillment. The context of a game often provides a much more consistent system of reward, progression, and success, than many other contexts that people experience in their daily lives. This feature means that some people may be more motivated to succeed in the predictable context of a game compared to another context in their life, but also that a game can be intrinsically motivating.
>
> While a reward or progression is almost always a motivating factor, motivation does not have to be external to the player. An internal feeling of "flow" can also be a rewarding experience which motivates the person to continue playing. A "flow" state happens

> when a person is fully engaged in a task which requires a skill, has clear goals and constant feedback, and where the person has control and concentration. It is important that the task is neither too easy or difficult.
>
> There are also more specific and personal motivations for playing games. Some people may be motivated by the chance to break away from social norms and their own identity and act as another character. Other people may be motivated to play a game that allows them to socialize with people who align with their personality and values. Other people may be motivated by the opportunity to practice and improve at a skill.

Saltzman (2022) considers game playing by older adults.

> More players over the age of 65 say they play to "use my brain" than any other age group (68%). For men, two-thirds also play to have fun (67%) and pass time (66%), a similar number of women also play to pass time (70%) and to unwind and relax (66%). Other popular reasons are that video games bring joy (93%), provide mental stimulation (91%) and stress relief (89%).
>
> Particularly with the isolation brought on by the pandemic, people connected through games and enjoyed shared experiences. In 2020, 65% of people said they played together, which jumped to 77% in 2021 and now it's at an incredible 83% who play together. Games have become a great way to convene with people socially and to connect with family and friends as well as the global community.

Vlachopoulos and Makri (2017) consider the effect of games and simulations on higher education. Their systematic literature review begins with a thorough review of earlier reviews. They then focus on 123 studies relevant to the effect of games and simulations on higher education. They organize their findings into three categories: cognitive, behavioral, and affective outcomes.

> Cognitive outcomes refer to the knowledge structures relevant to perceiving games as artefacts for linking knowledge-oriented activities with cognitive outcomes. Serious gaming, especially given the context of enthusiastic students, has proved to be an effective training method in domains such as medical education, for example, in clinical decision-making and patient interaction.

They also address how "virtual simulations can enhance complex cognitive skills, such as self-assessment" Chapters 7 and 8 delve into these phenomena.

> Behavioral objectives for higher education students refer to the enhancement of teamwork and improvement in relational abilities, as well as stronger organizational skills, adaptability and the ability to resolve conflicts. Simulation games are often seen as powerful tools in promoting teamwork and team dynamics, collaboration, social and emotional skills, and other soft skills, including project management, self-reflection, and leadership skills, which are acquired through reality-based scenarios with action-oriented activities.
>
> Affective outcomes of using games and simulations in the learning process address student engagement, motivation, and satisfaction.

In this and the other two categories, they observe that "There seems to be a lack of shared definitions or taxonomy necessary for a common classification, which, therefore, results in terminological ambiguity."

Van Dijk and De Dreu (2021) argue that "Games are a powerful tool to identify the neural and psychological mechanisms underlying interpersonal and group cooperation and coordination." They report that

> Important advances have been made in uncovering the neurobiological underpinnings of key factors involved in cooperation and coordination, including social preferences, cooperative beliefs, (emotion) signaling, and, in particular, reputations and reciprocity. This had led to an increased focus on group heterogeneities, intergroup polarization and conflict, cross-cultural differences in cooperation and norm enforcement, and neurocomputational modeling of the formation and updating of social preferences and beliefs.

Games for K–12

In a recent study (Rouse, Lombardi, & Gargano, 2023), we found that games, as well as camps and projects, that provide well-designed, engaging student experiences provide at least moderate positive impacts on:

- Knowledge & Skills Gained
- Personal Ability Benefits
- Personal Attitude Benefits

Figure 4.1 summarizes the relationships among these outcomes. Attitude in terms of motivation, engagement, and self-perceptions is the key to fostering abilities and knowledge and skills. Interventions need to be sufficiently compelling to capture students' interests in ways that they feel empowered to act.

Engagement can be sustained for durations of a week or so and have greater impacts than longer interventions that lose their novelty. However, games with multiple levels can sustain interest

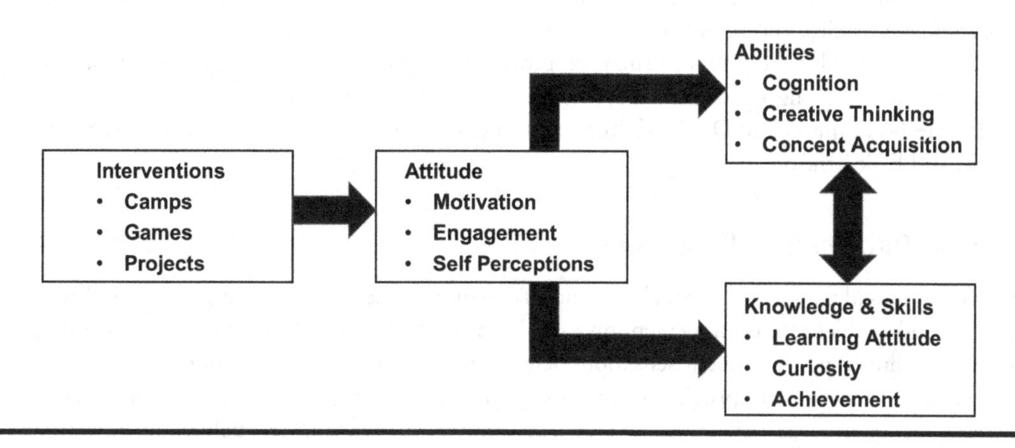

Figure 4.1 Relationships among interventions, attitude, and abilities.

if carefully designed and evaluated. Successive levels need to unveil greater challenges, more information, and additional tools.

For the games reviewed, there were limited gender differences as the interventions were designed with such possibilities in mind. It appears that gender differences emerge in environments where the designers were not sensitive to these phenomena, e.g., by designing all the characters as males.

Observations on Games

- These observations are based on several hundred studies of the efficacy of games for K–12 education, including several meta-analyses
- Games are more likely to contribute to achieving educational objectives if they are specifically designed to meet these objectives
- For older students, games often involve interactive, and possibly immersive, environments to support problem understanding and solution design
- For younger students, games are usually board games, often computer-generated, or videos that involve quizzes to prompt student understanding of content

Summary

Motivation, engagement, and self-perceptions of efficacy are essential for game players to invest their interest and energy in earnestly playing games. If they just "go through the motions," the outcomes for which the games were designed and developed are unlikely to be achieved.

Learning

Figure 4.2 summarizes the primary learning objectives for 30 games in Chapter 2. This tabulation resulted when the entries in Tables 2.1 to 2.4 were aggregated, classified, sorted, reconsidered, and resorted. This, admittedly, required fairly detailed knowledge of these 30 games, far beyond what is captured in these four tables.

Table 4.1 provides definitions of the aggregated learning objectives summarized in Figure 4.2. In this section, I consider how these objectives can be addressed in terms of human psychological processes, tendencies, and preferences.

I have adopted a human information processing perspective for addressing these issues. Therefore, I am relying on cognitive science and cognitive psychology, as pioneered by George Miller, Herbert Simon, and Daniel Kahneman rather than, for example, William James, Sigmund Freud, and Erik Erikson.

Human Information Processing

Figure 4.3 provides a useful approach to conceptualizing human information processing. Memory informs and enables sensation, perception, cognition, and actuation. These processes contribute to memory, influencing subsequent sensation, perception, cognition, and actuation.

Memory is often conceptualized in terms of short-term, long-term, etc. A subset of memory includes mental models. This ambitious construct embraces a broad range of behavioral phenomena. It also has prompted various controversies. On one hand, it would seem that people must have

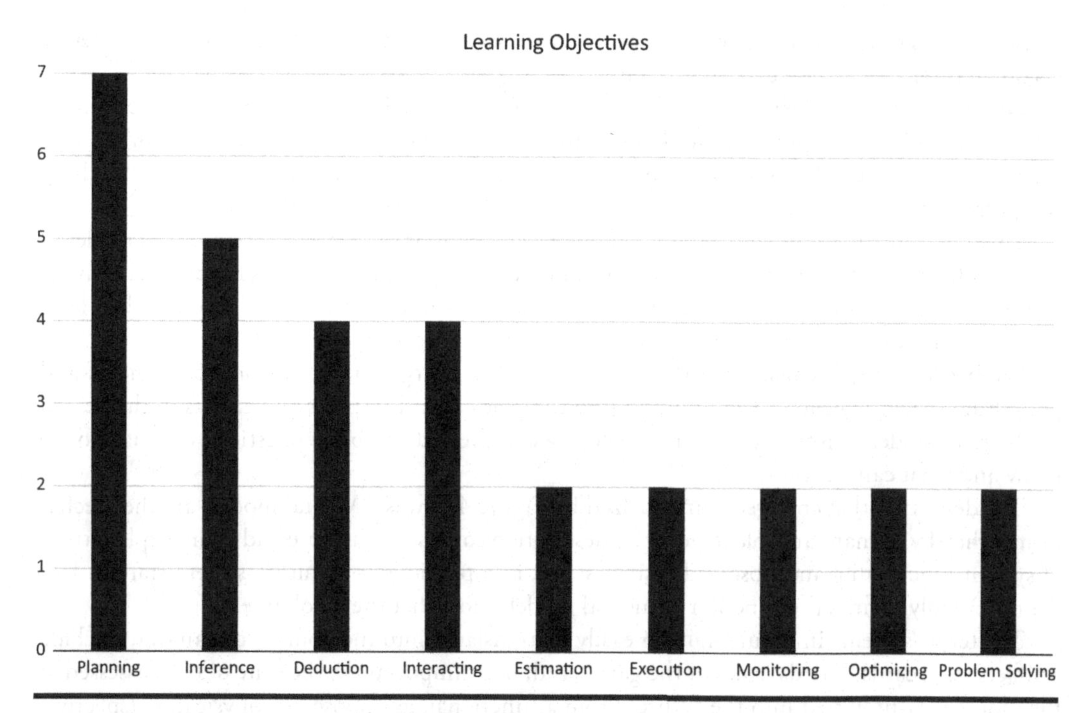

Figure 4.2 Primary learning objectives for 30 games in Chapter 2.

Table 4.1 Definitions of Aggregated Learning Objectives

Learning Objective	*Definition*
Planning	Selecting objectives, formulating strategies, and developing plans to achieve chosen objectives
Inference	Reaching a conclusion on the basis of evidence and reasoning
Deduction	Using understanding of underlying mechanisms, general laws, and principles to reach conclusions
Interacting	Developing an understanding of stakeholders' values, concerns, and perceptions, as well as their implications
Estimation	Prediction, filtering, and smoothing of data, which may be ad hoc, to characterize the assumed underlying phenomenon
Execution	Implementing courses of action associated with given plans to achieve desired outcomes
Monitoring	Observing ongoing actions and outcomes to infer and/or deduce acceptability and appropriateness of activities
Optimizing	Devising courses of action that maximize, e.g., profits, or minimize, e.g., costs, desired outcomes
Problem Solving	Detecting, diagnosing, compensating, and remediating undesirable states and outcomes

mental models of their cars, for example. Otherwise, how would people be able to so proficiently negotiate traffic, park their cars, and so on?

On the other hand, perhaps people have just stored in their memories a large repertoire of patterns of their car's input–output characteristics, a large look-up table, if you will, of steering wheel angles and vehicle responses. From this perspective, there is no "model" per se – nothing computational that derives steering wheel angle from desired position of the car.

This is a difficult argument to resolve if one needs proof of the representation of the model in one's head. Are their differential equations, neural nets, or rules in the driver's head? Alternatively, one might adopt a functional point of view and simply claim that humans act as if they have certain forms of model in their brains that enable particular classes of behavior.

We became deeply engaged in this issue, reviewing a large number of previous studies and publishing the highly cited, "On Looking Into the Black Box: Prospects and Limits in the Search for Mental Models" (Rouse & Morris, 1986). We addressed the basic questions of what do we know and what can we know.

The definition that emerged, summarized by Figure 4.4, was, "Mental models are the mechanisms whereby humans are able to generate descriptions of system purpose and form, explanations of system functioning and observed system states, and predictions of future system states." This definition only defines the function of mental models, not what they look like.

The term "system" in Figure 4.4 can easily be translated into the domain of games to include artifacts, e.g., cards, boards, rules of the game, and, most importantly, the state of play, assessed at the moment or projected into the future. From a functional, as opposed to physical, perspective, the elements of Figure 4.4 are central to the psychology of games.

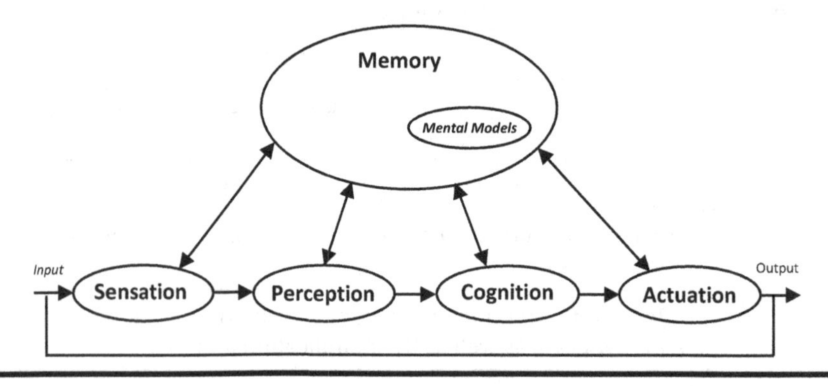

Figure 4.3 Human information processing.

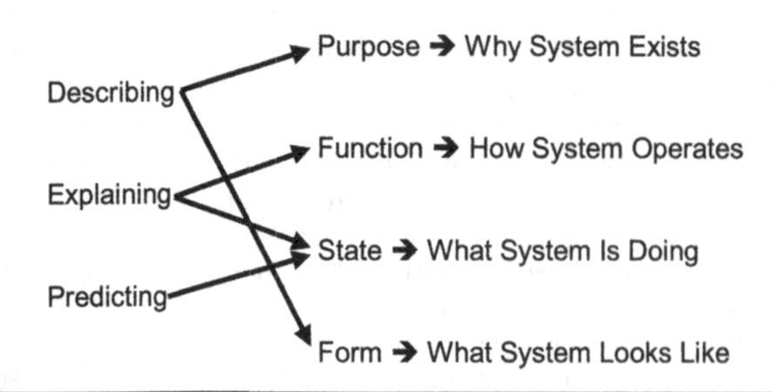

Figure 4.4 Functions of mental models.

People also have mental models of other people, including team members and competitors. We hypothesized that shared models of teamwork are central to team performance (Rouse, Cannon-Bowers, & Salas, 1992). Our definition of mental models for teamwork followed the functional definition in Figure 4.4. However, the knowledge content, as shown in Figure 4.5, differs from our earlier discussions.

Our research showed that team members need to know what is expected of them by other team members and know what they could expect of others. Without accurate expectations, people might not communicate or communicate erroneously or ambiguously. This can result in game players misinterpreting the actions and intentions of team members and competitors. Improved shared mental models of teamwork led to the creation of Team Model Training (TMT) that was discussed in Chapter 2.

We also applied this thinking to a study of teamwork in the performing arts (Rouse & Rouse, 2004). This interview study involved 12 performing arts including straight and improv theater, symphony and jazz, ballet, etc. Using context-specific versions of Figure 4.5, we devised a framework for assessing the extent to which team training is utilized in each art. We found that coordination and training are strongly influenced by the role of the leader and the familiarity of team members, which varied across the 12 performing arts studied. For example, if the leader prepares the team but does not participate in the performance, coordination is much more explicitly programmed.

Estimation and Decision Making

The cognition of estimation and decision making is of central importance to understanding the psychology of games. Estimation can involve deciding what did happen, inferred from data, or deciding what is likely to happen, perhaps deduced from one's mental models.

Quite some time ago, I studied people's abilities to predict future points in dynamic time series, or smooth past points of such time series. The experimental task was akin to air traffic control where subjects were asked to predict where an aircraft would next be. I employed stochastic estimation theory to model subjects' formation of mental models of the time series. We found

Level	Types of Knowledge		
	What	How	Why
Detailed/ Specific/ Concrete	Roles of Team Members (Who Member Is)	Functioning of Team Members (How Member Performs)	Requirements Fulfilled (Why Member Is Needed)
↓	Relationships Among Team Members (Who Relates to Who)	Co-Functioning of Team Members (How Members Perform Together)	Objectives Supported (Why Team Is Needed)
Global/ General/ Abstract	Temporal Patterns of Team Performance (What Typically Happens)	Overall Mechanisms of Team Performance (How Performance Is Accomplished)	Behavioral Principles/Theories (Why: Psychology, Management, Etc.)

Figure 4.5 Knowledge content of mental models for teamwork.

that subjects discounted past points of the time series too heavily, leading to less accurate mental models and poorer predictions (Rouse, 1977).

Human decision making is not always as crisply rational as stochastic estimation theory. Humans tend to be flawed decision makers. Herbert Simon (1957, 1969, 1972) pioneered studies of bounded rationality and satisficing by humans, i.e., pursuing satisfactory rather than optimal decisions. George Miller's most famous discovery was that human short-term memory is generally limited to holding seven pieces of information, plus or minus two (Miller, 1956).

Ward Edwards (1967) found people have difficulty processing probabilistic information. Daniel Kahneman and Amos Tversky explored the heuristics and biases that limit effective decision making (Kahneman, 2011). Biases include confirmation bias (only taking into account information that supports presuppositions), optimism bias (over-estimating the likelihood of success), availability bias (over-estimating the frequency of familiar events), and fundamental attribution error (inferring incorrect causation).

Henry Mintzberg (1975) researched the folklore of management decision making and found the belief that managers are reflective, careful optimizers is unfounded. Instead, typical managers are interrupt-driven, spending a few minutes on each task. Gary Klein (1998, 2003, 2004) found that when professionals are faced with decisions they have addressed many times, their intuitions are often correct unless, of course, their perception that this is a familiar situation is unwarranted.

Kahneman and Richard Thaler have been pioneers in the field known as behavioral economics. Thaler's study of nudges (2009, 2015) has illustrated the subtlety of what may affect decisions. This integration of psychology into models of people understanding and playing games enables richer insights into the decisions they consider and adopt. It makes no sense to assume that people will behave in ways that are unlikely or perhaps impossible.

The training and aiding of human decision making can make a huge difference (Rouse, 2019). Training increases people's potential to perform. Aiding directly augments human performance. Thus, we can compensate for humans' limitations by enhancing abilities, overcoming limitations, and fostering acceptance of such assistance. I address this in a later section of this chapter.

Planning and Problem Solving

Early studies of human problem solving began in the 1930s. Research in the 1960s and 1970s focused on simple, albeit novel, laboratory tasks, e.g., the game *Tower of Hanoi*. Allen Newell and Herbert Simon's *Human Problem Solving* (1972) is seen as a classic resulting from such studies. Their rule-based paradigm, which they termed production systems, became a standard in cognitive science.

More recently, greater emphasis has been placed on the study of real problem solvers performing real-world tasks. Research on detection and diagnosis of systems failures (Rasmussen & Rouse, 1981; Rouse, 2007) is a good example. Rouse and Spohrer (2018) address situations where artificial intelligence can be used to augment human problem solving in domains ranging from driverless cars to healthcare to insurance underwriting.

Typical assumptions associated with models based on problem-solving theory include a specified human mental model of the problem domain and known information utilization behaviors, a repertoire of symptom patterns, and troubleshooting rules. The phenomena usually predicted by such models include time until the problem is solved, steps until the problem is solved, and the frequency of problem-solving errors.

Drawing upon a wide range of sources, I developed a general three-level representation of human problem solving (Rouse, 1983, 2007). Rasmussen's distinctions among skill-based, rule-based, and knowledge-based behaviors (Rasmussen, 1983), in combination with Newell and Simon's (1972) theory of human problem solving, led to the conclusion that problem solving occurs on more than one level – see Figure 4.6.

When humans encounter a decision-making or problem-solving situation, they usually have some expectations associated with the context of the situation. They perhaps unconsciously invoke a frame (Minsky, 1975) associated with this situation. Frames describe what one can expect in an archetypal situation, e.g., a wedding versus a funeral.

Based on the frame, they then consider available information on the state of the system. If this information maps to a familiar pattern, whether normal or abnormal, it enables them to activate scripts (Schank & Abelson, 1977) that enable them to act, perhaps immediately, via recognition-primed symptomatic rules (S-Rules) that guide their behaviors (Klein, 2004).

If the observed pattern of state information does not map to a familiar pattern, humans must resort to conscious problem solving and planning (Johannsen & Rouse, 1983), perhaps via analogies or even basic principles. Based on the structure of the problem, which typically involves much more than solely observed state variables, they formulate a plan of action and then execute the plan via structural topographic rules (T-Rules). As this process proceeds, they may encounter familiar patterns at a deeper level of the problem and revert to relevant S-Rules.

It often does not make sense to represent human behavior and performance for any particular task using only one type of model. Scripted behaviors may be reasonable for familiar and frequent instances of these tasks. However, for unfamiliar and/or infrequent instances of these tasks, a more robust representation is likely to be needed. The notion of S-Rules and T-Rules supports this distinction. S-Rules are very powerful. Patterns of cues provide prompt immediate, and occasionally erroneous, recognition of frames, scripts, and actions. T-Rules are less efficient but, when needed, will lead to successful problem solving.

Training and Aiding

A primary purpose of games is for players to enjoy these games and have fun. In some situations, an equally important objective is for players to gain knowledge and skills that improve their game outcomes and transfer to related tasks, perhaps associated with their employment. This was the primary objective of the simulation games summarized in Table 2.4.

In terms of the foregoing discussion of mental models, estimation and decision making, and planning and problem solving, a central question is what mental models should we attempt to

	Decision	**State-Oriented Response**	**Structure-Oriented Response**
Recognition & Classification	Frame Available?	Invoke Frame	Use Analogy and/or Basic Principles
Planning	Script Available?	Invoke Script	Formulate Plan
Execution & Monitoring	Pattern Familiar?	Apply Appropriate S-Rules	Apply Appropriate T-Rules

Figure 4.6 Problem solving, decisions, and responses.

create and how should they be fostered? Do people need to know theories, fundamentals, and principles, or can they just learn practices? The significance of this question cannot be overstated. I introduce this topic here, and explore it in more detail in Chapters 7 and 8.

There are several issues to address. Should we increase players' potential to perform via training or directly augment performance via aiding? Should we primarily focus on improving players' game performance or be more concerned with improving players' knowledge and skills? What mechanisms best support transfer of learning from games to actual work?

Training

Should players learn about how the system works or how to work the system? Is the system the game, or is the game a means to gain knowledge and skills applicable to other tasks? The former implies gaining process understanding, while the latter is more concerned with understanding procedures and practicing them. Driving a car does not require understanding how the drive train functions; maintaining a car does.

Often, a range of training interventions is available. Sometimes a mix of interventions makes great sense. For example, our experiences with a range of training endeavors have made it clear that immersing novice trainees in full-scope training simulations can often be overwhelming. Somewhat simplistically, it makes great sense to have students learn how to play Rummy before they tackle Bridge or Cribbage.

The concept of mixed-fidelity training, shown in Figure 4.7, emerged. Put simply, the low-fidelity simulator, and associated instruction, prepares trainees to perform in the moderate-fidelity simulator, which in turn prepares them to perform in the high-fidelity simulator (Rouse, 1982–1983).

Beyond the pedagogical value of this approach, it also better manages the flow of trainees. The low and moderate-fidelity simulators were typically hosted on personal computers. Thus, for

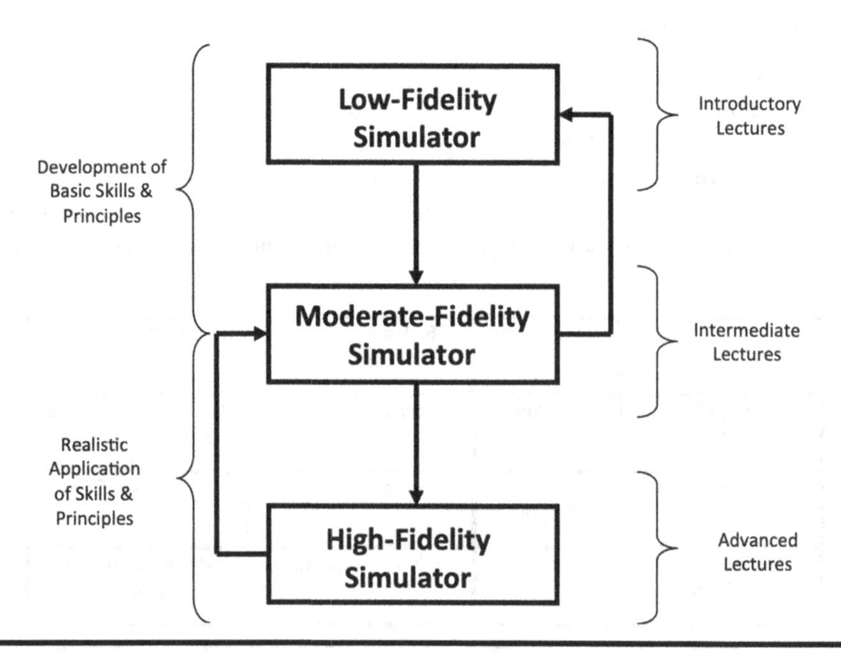

Figure 4.7 Mixed-fidelity training.

example, one could have a dozen trainees using these simulators while one or two were using the high-fidelity simulator.

This was the case for several of our training efforts. TASK and FAULT from Table 2.4 served as the low- and moderate-fidelity simulators, respectively, to support Marine Safety International's high fidelity supertanker engine room simulator, which only allowed for one trainee at a time. TMT, also from Table 2.4, supported trainees prior to their engagement with the high fidelity, full scope simulator of the Aegis Combat Information Center, which involved the full complement of 25 Navy crew members.

I address the need to integrate multiple training interventions in more detail in Chapters 7 and 8. A key concern is the extent to which learning objectives are primarily related to game performance versus the game being a means to enhance the performance of other tasks.

For example, is the purpose of learning calculus to enhance framing and solving calculus problems, or is the intent also for students to learn how calculus can be applied to addressing important engineering problems? I expect that the transfer of calculus knowledge and skills to solving engineering problems is often an important objective.

Aiding

Aiding involves directly augmenting people's performance. Aiding can help players with game performance, or it may involve augmenting the performance of tasks that the game was intended to train people to subsequently perform.

Procedures are a form of job aid, often valuable during emergency situations. Aids for procedure monitoring support humans by inferring errors of omission and commission. The results are error-tolerant interfaces that I later discuss.

Aiding such as bidding rules for Bridge provide guidance on how to value your own hand and infer your partner's hand. Easley Blackwood proposed a slam-seeking convention in 1933, slams involving winning all 13 tricks. It gained awareness and is used by many seasoned players and is referred to as the Blackwood Slam Convention.

Projections of the impacts of one's own candidate game moves can be quite helpful. Inferring opponents' intentions, strategies, and plans can inform such choices. Bayesian probability estimation has been applied to making such inferences. Memory aids for past moves by self and others can help by, for example, keeping track of what cards are left to be played.

For games – or jobs – that involve problem solving in networked domains, symptom tracing can decrease time to solutions and reduce erroneous test choices. For example, portraying information on what is still working can substantially reduce the size of the feasible set that needs to be explored. Such environments are often amenable to guidance on half-split tests that halve the feasible set with each test.

Intelligent Interfaces

How might artificial intelligence (AI) enhance aiding? There are considerable concerns that AI will take over a wide range of jobs, including playing your opponents in online games. However, in many situations, AI will be used to augment human intelligence, rather than being deployed to automate intelligence and replace humans (Rouse & Spohrer, 2018). What functions are needed to augment intelligence?

Information Management. One function will be information management (Rouse, 2007, 2019). This involves information selection (what to present) and scheduling (when to present it).

Information modality selection involves choosing among visual, auditory, and tactile channels. Information formatting concerns choosing the best levels of abstraction (concept) and aggregation (detail) for the tasks at hand. AI can be used to make all these choices in real time as the human is pursuing the tasks of interest.

Intent Inferencing. Another function is intent inferencing (Rouse, 2007, 2019). Information management can be more helpful if it knows both what humans are doing and what they intend to do. Representing humans' task structures in terms of goals, plans, and scripts (Schank & Abelson, 1977) can enable making such inferences. Scripts are sequences of actions to which are connected information and control requirements. When the intelligence infers what you intend to do, it then knows what information you need and what controls you want to execute it.

One of the reasons that humans are often included in systems is because they can deal with ambiguity and figure out what to do. Occasionally, what they decide to do has potentially unfortunate consequences. In such cases, "human errors" are reported. Errors in themselves are not the problem. The consequences are the problem.

Error-Tolerant Interfaces. For this reason, another function is an error-tolerant interface (Rouse & Morris, 1987; Rouse, 2007, 2019). This requires capabilities to identify and classify errors, which are defined as actions that do not make sense (commissions) or the lack of actions (omissions) that seem warranted at the time. Identification and classification lead to remediation. This occurs at three levels: monitoring, feedback, and control. Monitoring involves collection of more evidence to support the error assessment. Feedback involves making sure the humans realize what they just did. This usually results in humans immediately correcting their errors. Control involves the automation taking over, e.g., applying the brakes, to avoid the imminent consequences.

Adaptive Aiding. The notion of taking control raises the overall issue of whether humans or computers should perform particular tasks. There are many cases where the answer is situation dependent. Thus, this function is termed adaptive aiding (Rouse, 1988, 2007, 2019). The overall concept is to have mechanisms that enable real-time determination of who should be in control. Such mechanisms have been researched extensively, resulting in a framework for design that includes principles of adaptation and principles of interaction. A First Law of Adaptive Aiding has been proposed – computers can take tasks, but they cannot give them.

Intelligent Tutoring. Another function is intelligent tutoring to both train humans and keep them sufficiently in the loop to enable successful human task performance when needed. As noted earlier, training usually addresses two questions: 1) how the system works and, 2) how to work the system. Keeping humans in the loop addresses maintaining competence. Unless tasks can be automated to perfection, humans' competencies need to be maintained. Not surprisingly, this often results in training versus aiding tradeoffs, for which guidance has been developed (Rouse, 2007, 2019).

How do all of the above functions fit together to create an intelligent interface? How will players of games and performers of jobs interact with such capabilities? This relates to the future of gaming. I address these questions in Chapter 10.

Conclusions

This chapter has illustrated the breadth with which this book addresses the construct of "games." The psychology of games concerns motivation and engagement as well as behavior and performance in game play. The games of interest include games for fun, games for learning, and games for job training. The players range from K–12 students to sports fans, to supertanker chief engineers to Navy combat crews.

We began by considering the psychology of engagement. I briefly reviewed the rich research literature on what motivates people to engage in game playing. It certainly is multi-dimensional. I summarized our recent findings on engagement of K–12 students in games as it affects knowledge and skills gained and consequent abilities. Engagement is a prerequisite to learning.

An analysis of the learning objectives associated with the 30 games summarized in Chapter 2 resulted in identifying nine aggregate objectives. Planning predominates, followed by inference, deduction, and interacting. Interacting involves understanding other people, whether they be partners, team members, competitors, or adversaries.

To achieve these learning objectives, we need to understand the human information processing that enables playing games. I articulated an overall model involving sensation, perception, cognition, and actuation, all enabled by memory as well as contributing to memory. I reviewed the state of knowledge on human mental models, estimation and decision making, and planning and problem solving. These constructs should inform game design and evaluation.

Training and aiding are two types of interventions intended to create the potential for humans to perform, and directly augment human performance, respectively. The objectives include enabling people to successfully play games, and transfer the knowledge and skills gained to other tasks, perhaps associated with their work. I introduced the notion of intelligent interfaces and outlined the functions this construct is intended to provide.

Chapters 5 and 6 address the sociology and politics of games, respectively. This requires that we understand the ecosystems within which games are played. I move beyond cards, boards, and screens to larger human relationships to consider pervasive societal phenomena and impacts.

Chapters 7 and 8 address games for learning and serious games, respectively. Such games are intended to be fun and engaging, but the overarching objective involves practical outcomes for students, employees, and people in general. These chapters draw upon much of the material reviewed in this chapter on the psychology of games.

Chapters 9 and 10 consider the games industry and the future of gaming, respectively. The industry is large and growing strongly, enabled by technology trends and motivated by attractive economic returns. Another driving force is the increasing breadth of applications of games, catalyzed in part by the coronavirus pandemic where we learned how many more activities could be successfully conducted online.

I consider the future of gaming as it provides an opportunity to project possible futures of technology-enabled games. I consider likely impacts on health, education, and energy, as well as civic realms of government and politics. Games are very likely to become increasingly pervasive and powerful.

References

Acaster, S. (2023). *The Psychology of Gaming – Why Do People Play Games? Outschool. https://info.outschool.com/uk-blog/the-psychology-of-gaming-why-do-people-play-games.*

Edwards, W., & Tversky, A. (Eds.). (1967). *Decision Making.* New York: Penguin.

Johannsen, G., & Rouse, W.B. (1983). Studies of planning behavior of aircraft pilots in normal, abnormal, and emergency situations. *IEEE Transactions on Systems, Man and Cybernetics, SMC-13,* (3), 267–278.

Kahneman, D. (2011). *Thinking, Fast and Slow.* New York: Farrar, Straus and Giroux.

Klein, G. (1998). *Sources of Power: How People Make Decisions.* Cambridge, MA: MIT Press.

Klein, G. (2003). *Intuition at Work: Why Developing Your Gut Instincts Will Make You Better at What You Do.* New York: Doubleday.

Klein, G. (2004). *The Power of Intuition: How to Use Your Gut Feelings to Make Better Decisions at Work.* New York: Currency.
Klein, G. (2004). *The Power of Intuition: How to Use Your Gut Feelings to Make Better Decisions at Work.* New York: Currency.
Miller, G.A. (1956). The magical number seven, plus or minus two: Some limits on our capacity for processing information. *Psychological Review*, 63 (2), 81–97.

Minsky, M. (1975). A framework for representing knowledge. In P.H. Winston, Ed., *The Psychology of Computer Vision.* New York: McGraw-Hill.

Mintzberg, H. (1975). The manager's job: Folklore and fact. *Harvard Business Review*, July–August.

Newell, A., & Simon, H.A. (1972). *Human Problem Solving.* Englewood Cliffs, NJ: Prentice-Hall.

Rasmussen, J. (1983). Skills, rules, and knowledge; signals, signs, and symbols, and other distinctions in human performance models. *IEEE Transactions on Systems, Man, & Cybernetics, SMC-13,* (3), 257–266. https://doi.org/10.1109/TSMC.1983.6313160.

Rasmussen, J., & Rouse, W.B. (Eds.). (1981). *Human Detection and Diagnosis of System Failures.* New York: Plenum Press.

Rouse, W.B. (1977). A theory of human decision making in stochastic estimation tasks. *IEEE Transactions on Systems, Man, and Cybernetics, SMC-7,* (3), 274–283.

Rouse, W.B. (1982–1983). A mixed-fidelity approach to technical training. *Journal of Educational Technology Systems,* 11 (2), 103–115.

Rouse, W.B. (1983). Models of human problem solving: Detection, diagnosis, and compensation for system failures. *Automatica*, 19 (6), 613–625.

Rouse, W.B. (1988). Adaptive aiding for human/computer control. *Human Factors*, 30 (4), 431–443.

Rouse, W.B. (2007). *People and Organization: Explorations of Human-Centered Design.* New York: Wiley.

Rouse, W.B. (2019). *Computing Possible Futures: Model Based Explorations of "What if?"* Oxford, UK: Oxford University Press.

Rouse, W.B., & Morris, N.M. (1986). On looking into the black box: Prospects and limits in the search for mental models. *Psychological Bulletin*, 100 (3), 349–363.

Rouse, W.B., & Morris, N.M. (1987). Conceptual design of a human error tolerant interface for complex engineering systems. *Automatica*, 23 (2), 231–235.

Rouse, W.B., & Rouse, R.K. (2004). Teamwork in the performing arts. *Proceedings of the IEEE*, 92 (4), 606–615.

Rouse, W.B., & Spohrer, J.C. (2018). Automating versus augmenting intelligence. *Journal of Enterprise Transformation.* https://doi.org/10.1080/19488289.2018.1424059.

Rouse, W.B., Cannon-Bowers, J.A., & Salas, E. (1992). The role of mental models in team performance in complex systems. *IEEE Transactions on Systems, Man, and Cybernetics*, 22 (6), 1296–1307.

Rouse, W.B., Lombardi, J.V., & Gargano, M. (2023). *Policy Innovations to Enhance the STEM Talent Pipeline: Interventions to Increase STEM Readiness of K-12 Students.* Hoboken, NJ: Systems Engineering Research Center, Stevens Institute of Technology.

Ryan, R.M., & Deci, E.L. (2000). Intrinsic and extrinsic motivations: Classic definitions and new directions. *Contemporary Educational Psychology*, 25 (1), 54–67.

Ryan, R.M., Rigby, C.R., & Przybylski, A. (2006). The motivational pull of video games: A self-determination theory approach. *Motivation & Emotion*, 30 (4), 344–360.

Saltzman, M. (2022). More adults play video games than kids – and more surprising stats. *USA Today*, June 11.

Schank, R., & Abelson, R. P. (1977). *Scripts, Plans, Goals and Understanding: An Inquiry into Human Knowledge Structures.* Hillsdale, NJ: Erlbaum.

Simon, H.A. (1957). *Models of Man: Social and Rational.* New York: Wiley.

Simon, H.A. (1969). *The Sciences of the Artificial.* Cambridge, MA: MIT Press.

Simon, H.A. (1972). Theories of bounded rationality. In C.B. McGuire & R. Radner, Eds., *Decision and Organization* (Chap. 8). New York: North Holland.

Thaler, R.H. (2015). *Misbehaving: The Making of Behavioral Economics.* New York: Norton.

Thaler, R.H., & Sunstein, C.R. (2009). *Nudge: Improving Decisions About Health, Wealth, and Happiness.* New York: Penguin.

Trevisan, J.P. (2023). *The Psychology of Motivation in Games*. Vila Madalena, Brazil: Escola Britânica de Artes Criativas.

Van Dijk, E., & De Dreu, C.K.W. (2021). Experimental Games and Social Decision Making. *Annual Review of Psychology*, 72, 415–438.

Vlachopoulos, D., & Makri, A. (2017). The effect of games and simulations on higher education: A systematic literature review. *Technology in Higher Education*, 14 (22). https://doi.org/10.1186/s41239-017-0062-1.

Von Ehren, S. (2020). Why do people love games? *The New York Times*, June 11.

Chapter 5

Sociology of Games

Introduction

The sociology of games considers human–human interactions in the context of game play, as well as off the field. The first half of this chapter addresses the sociology of sports games, which play central roles in many societies. Games as social outlets are an important consideration, particularly in terms of social affiliations with sports teams. The second half of this chapter concerns the sociology of online games. Important considerations include how teams of players come together and players' relationships beyond the games.

The content of this chapter is a mix of social psychology and sociology. From Chapter 4, we understand individual game performance pretty well. Social interactions and behaviors are much more complicated, especially for large social networks, which social media has enormously enlarged. Social psychology considers people and organizations in relation to each other. There are several aspects of social psychology that are of particular interest. A central issue is the extent to which people trust information provided by others and, in particular, trust others' intentions.

Sociology attempts to explain social trends and events as they affect societal changes over time, often over long periods of time. A good example is the transition from agrarian to manufacturing to service economies. More specifically, this discipline focuses on explaining how we arrived at a current social situation, e.g., welfare state. They seldom address trying to predict the efficacy of near-term interventions. In the context of this chapter, the societal role and evolution of sports fits this paradigm.

Sociology of Sports

The sociology of sport can be approached from several perspectives (Wikipedia, 2023).

- Functionalism: Society is a complex system whose parts work together to promote solidarity and stability. Relates to sport and religious ceremonies. Both create a sense of community.
- Interpretative Sociology: Relations of social action to status, subjectivity, meaning, motives, identities, and social change. Sports have been secularized and involve a meritocracy and specialization.

DOI: 10.4324/9781003491927-5

- Neo-Marxism: Sport as an ideological tool of the bourgeoisie used to deceive the masses in order to maintain control.
- Cultural studies: Hegemony (dominance) addresses the relations of power, as well as methods and techniques used by dominant groups to achieve ideological consent, without resorting to physical coercion.

> Sport sociology is the study of the relationship between sport and society. It examines how culture and values influence sport, how sport influences culture and values, and the relationship between sport and the major social spheres of life such as the media, politics, the economy, religion, race, gender and youth.
>
> **(Serra, 2015)**

> Sport, at a social, recreational or competitive elite level, can be considered as a microcosm, or small-scale version, of society. The same social issues that exist in larger society also exist in sport. These social issues, which can be seen in larger society and also in sport, are the concepts of values, race, gender, ethnicity, class, sexuality, youth, politics, religion and economics.
>
> When we consider sport as a microcosm of society and/or we determine that sport teams mirror society, we see an important likeness between the two and must thus engage with social issues such as racism, discrimination, inequalities and homophobia that are revealed in sport and society. The sociology of sport uses essential and conflicting approaches that force us to explore alternative ways of viewing the place and organization of sport in our society and how issues and problems presented by sport in society affect individuals.

Broch (2022) observes,

> Sports, with their familiar seasonal patterns, are created and recreated as cultural systems gravitationally bound by our play to familiar symbols, myth, codes, and narratives. In their chase for records, statistical measurements, and "fair" crowning of a one true champion, sports can come off as the epitome of rational and bureaucratic modernity.
>
> In various ways, sports fuse athletic and social performances. Athletes execute left and right turns, jumps, pushes, and pulls, as well as throws, catches and pirouettes. At the same time, or better yet, prior to the moment of action, and in its aftermath, the experiential realm of athletic conquests is interpreted and imbued with codes, myths, and narratives. In this process, sport cultures themselves transform into symbols, metaphors, and background representations that we use to direct social life elsewhere.

Giulianotti and Thiel (2023) provide a very thoughtful treatise on how sociology, as an academic discipline, can contribute to research on the sociology of sports. Sociologists focus on

> de-constructing any errors, misunderstandings, inconsistencies, and contradictions that may be identified in the scientific, politic, medial, and public descriptions of

> social issues; examining the key features and patterns of social relations; comparing and contrasting, and identifying strengths and limitations, in theories, policies, and patterns of social relations; highlighting and investigating social relations of power, as characterized for example by social inequalities and divisions; and, identifying alternative possibilities for how societies may be organized, including within particular areas of social life, such as in sport.
>
> Sociology has consistently been an avant-garde discipline, in terms of identifying and highlighting progressive public issues that go on to gain some traction with wider publics, policy-makers, and corporations. Areas such as equality, diversity and inclusion and environment, social and governance—that are rooted in themes relating to social division and social justice, which have long been a major concern for sociologists and are illustrative of this avant-garde impulse.

Sports as Recreation

James Michener, a Pulitzer Prize-winning author, was a major advocate of sports as important recreation. He argues for lifelong involvement to maintain health and vitality. He outlines how and why people give up on exercise later in life, say beyond 30.

> As a nation we run the risk of forgetting the salutary effect of play. As adults we penalize ourselves unnecessarily by losing our capacity for it: the lazy flight of the frisbee with grown men and women chasing it ridiculously, running through the woods at a picnic, chasing with dogs over a freshly mown field, romping with kids on a lawn, playing stickball in the street, exhibiting lost prowess in a pick-up softball game, laughing with a beer can in the left hand while trying to toss a quoits ringer with the right.
>
> **(Michener, 1976)**

I have never been much of an athlete. Being tall, I played high school basketball but was a mediocre player. I never played baseball or football at any age. Growing up on an island in the Atlantic Ocean, I was often involved in sailing and boating in Rhode Island. Later in life, I took up running and participated in the Peachtree Road Race ten times and many other 10Ks for several years.

Later yet, I got hooked on hiking. We hiked the first 100 miles of the Appalachian Trail in Georgia and North Carolina several times. On one trip, we hiked to the top of Mt Elbert, the highest peak in the Colorado Rockies, and made it near to the peak of Mt Rainier in the Cascades. Severe back issues have resulted in my hiking now being limited to serious walking.

Sports as Public Entertainment

Michener (1976) notes that "All societies in all periods of history have needed some kind of public entertainment, and it has usually been provided by sports." He summarizes historical sports venues seen during his far-flung travels.

> Ancient Greece had its Olympiad and Rome its Colosseum. In the most distant corners of Asia Minor, I saw amphitheaters constructed by these civilizations because the rulers knew that the general citizenry required some kind of public entertainment. In

> Mérida, in western Spain, I visited the enormous flat plain that had once been walled to a height of four feet and waterproofed so that when a river was led into the area a small lake resulted on which actual ships could engage in simulated naval battles.
>
> In Crete young men and women skillfully leaped on and off the backs of charging bulls, and I have always been impressed by the frequency with which games are mentioned in the Bible. Some of the most effective analogies of St. Paul were borrowed from the arena. Shakespeare, too, found examples in sports, and I have found only one society in which sports were not a functional part. The Hebrews of Biblical time held a low opinion of games and said so, but when they entered Greek and Roman society they became advocates, like their neighbors.
>
> In all ages societies have looked to sports for entertainment, so that when the State of Alabama demands that its university provide first-class, big-time football, it is acting within a historical tradition, and when the State of Louisiana spends $163,000,000 to build a Superdome, it is aping only what Greece and Rome did ages ago. I am completely in favor of public sporting spectacles, for they fill a timeless need, but I am confused as to who should provide them and under what type of public sponsorship.

Michener indicates that no other country relies on higher education for public entertainment. With reference to football and basketball, "The American educational system has been called upon to provide enormously expensive training programs so that professional teams could prosper without putting up any money." In other words, higher education in the US subsidizes professional sports without compensation.

Branch (2011) observes,

> The United States is the only country in the world that hosts big-time sports at institutions of higher learning. College sports are deeply inscribed in the culture of our nation. Half a million young men and women play competitive intercollegiate sports each year. Millions of spectators flock into football stadiums each Saturday in the fall, and tens of millions more watch on television. The March Madness basketball tournament each spring has become a major national event, with upwards of 80 million watching it on television and talking about the games around the office water cooler. With so many people paying for tickets and watching on television, college sports has become Very Big Business.

Sports as Business

This overall economic situation has its merits. However, there is a central concern related to who benefits from all this business. Coaches and athletic directors at major programs make multi-million dollar salaries. Non-revenue sports and Title IX sports are heavily subsidized by the windfall from television revenues. Yet the basketball and football players, predominantly black, on whose performance the whole endeavor is based, earn nothing for their efforts.

Afshar (2015) characterizes the situation,

> A student-athlete is widely considered exactly what the name implies: a student first and an athlete second. The governing body of collegiate athletics—the National

> Collegiate Athletic Association (NCAA)—seeks to protect this status with strict rules that prevent these student-athletes from receiving compensation for their work or to have the right to leverage their personal brands, in a similar manner professional athletes do, without conceding their amateur status.
>
> Amateurism is a concept that contradicts an individual's publicity rights. Comparing the current NCAA model with the longstanding and continually evolving Olympics model, (one can) argue that the Olympics can serve as a sufficient example of granting athletes their rights of publicity while maintaining their status as amateurs.

This initial hurdle was overcome in 2021 with the NCAA "Name, Image, and Likeness" rule that grants athletes the right to benefit from their own publicity. The next, larger hurdle is allowing athletes to be paid for their performance. Various proposals are being discussed, with ardent opposition from traditionalists.

The total focus on money underlies the current realignment of college football conferences. Blackstone (2023) asserts that

> College football has been college football in name only for quite some time. In the past 20 to 25 years, it has been commercialized, commodified, capitalized and professionalized so much as to make it virtually indistinguishable from the NFL.
>
> This latest round of conference realignment has everything to do with chasing the billions of dollars broadcast corporations showered on the biggest schools the past couple of years—such as the $7 billion bag Fox, CBS, and NBC dropped on Big Ten members last summer to show their games—and nothing to do with mission statements such as, for example, Maryland's, which mentions athletics but once. Fact is, Maryland left the venerable ACC in 2014 after more than half a century for the riches the Big Ten could provide, to help soak up the debt on a football stadium it expanded without need.

Blackstone argues that universities should

> Pay and treat the football players as the employees they are, raking in hundreds of millions while making their coaches and athletic directors millionaires. Provide them the same health care and insurance as everyone else involved. And if they choose to matriculate at the school their team is attached to, let them do so like any other employee of the campus. Avail to them tuition remission.

Godfrey (2023) observes that

> Previous realignments were able to justify their true ends (money) by introducing consumer-friendly devices without the means looking quite as glaring. By overloading themselves for television revenue, power conferences are muddying their identity to the point that the game's premier franchises (sorry, schools) will find it logical in a few years to affiliate only with peers in the same financial standing, paying no concern to history or geography at all.
>
> That brand confusion will open cracks large enough for the game's biggest names to wriggle through—or rather, out of—their traditional conference affiliations when the

next big bang hits the sport after these current TV deals expire. The result will be the super league, something functionally and aesthetically closer to the NFL. This inevitable course will come at the expense of the regionality and tradition that made college football an idiosyncratic and uniquely American sport.

Professional Athletes

The NCAA insists that college basketball and football players are student-athletes, amateurs. However, the demands of major programs are such that these players should be characterized as unpaid professionals. Their tuition is waived, but many players, particularly blacks, will never get close to graduation.

Basketball players can enter the pro draft after one year. For football, it is three years. The chance of obtaining a pro contract is 1% in basketball and 2% in football. The average career in pro basketball is five years. For pro football, it is seven years, although running backs, for example, only average three years.

Basketball players' salaries are guaranteed but football players' salaries are contingent on sustained performance and not getting cut. Superstar football players can demand greater guarantees. Women's salaries, for instance in the Women's National Basketball Association, are much smaller than men's compensation.

Free agency is much better in basketball than football. Both leagues require four years of completed service. However, player mobility in the NBA is much greater. This is influenced by roster sizes. Basketball is 15 and football 55. Starting line-ups include 5 players in basketball and 22 players in football. Consequently, there are 450 pro basketball players and 1,696 pro football players.

The teams of the NBA generated combined revenues of US$10 billion in the 2021/22 season. The NFL teams generated combined revenues of US$12 billion. 21 of the 30 NBA teams were profitable. All 32 NFL teams were comfortably profitable. The Golden State Warriors is the most valuable NBA team at US$7.6 billion, while the Dallas Cowboys is the most valuable NFL team at US$9.2 billion.

I briefly mentioned the exploitation of black athletes in Chapter 2. This topic merits elaboration, drawing on Edwards (1969/2017), Michener (1976), Branch (2011), and Root (2013). The exploitation cuts across college and professional sports.

Many blacks see sports as a way out of poverty. Consequently, they devote substantial energies to becoming highly talented, particularly at basketball and football. People with these aspirations often devote less energy to academics. Nevertheless, their talents are sought by top college basketball and football programs.

Blacks constitute significant majorities on these high-revenue teams. Beyond not being compensated, as discussed earlier, many also do not get an education. Unprepared for college, many programs create easy, but useless, courses that enable maintaining acceptable grade point averages. I recall a final example question from one University of Georgia course: "How many points do you get for a three-point shot?"

A small percentage of these players make it to the pros. In football, they tend to play positions where injuries are common, e.g., running back. Thus, their careers are often short, e.g., average of three years. They may earn reasonable money for those few years, but not enough for the rest of their lives. They need jobs but lack education. One recent pro player admitted that he is illiterate, despite having spent time in college.

Edwards, Michener, and Branch argue that a portion of the immense revenues generated by these players should be used to ensure that they are educated and employable. This would also enable better decision making during the few years when they have substantial incomes.

Crime and Violence

Sports exist in the overall context of society. Consequently, the social and cultural phenomena prevalent in that context inevitably manifest themselves in sports. Organized crime has long been attracted to gambling, as well as other vices. Legalized vices do not deter criminals from attempting to secure a share of the resources involved.

Violence is another culturally prevalent phenomenon. Violence on the playing field is an inherent element of some sports, for example, American football. Violence initiated by spectators has long been common in some sports, for instance, soccer. The intensity of these games stimulates strong spectator engagement, including frustrations with losses and euphoria with victories.

Crime and Gambling. The United Nations Office on Drugs and Crime (2023) has recently reported on corruption in sport.

> Illegal betting and the related manipulation of sport competitions are major threats to the integrity of sport and to its nature. The role of illegal betting in sports in money-laundering has become a global problem and the financial scale of the problem is such that illegal betting is not only a major driver of corruption in sport, but also a major channel for money-laundering.
>
> While the clandestine nature of money-laundering makes it is difficult to estimate the amount of money that is laundered, the amount laundered globally in one year is estimated to be between $800 billion and $2 trillion, equal to between two and five per cent of global gross domestic product (GDP). Hence, government agencies, sports' governing bodies, and national and international sports organizations must take a coordinated approach to tackling these threats.

The European Data Journalism Network (2020) estimates that game fixing annually affects US$2 trillion globally. Football is most affected by game fixing, followed by tennis. During Covid, "ghost" matches were offered for betting. Fixers created non-existent line-ups, as well as match outcomes. Gamblers continued betting, unaware of such fixes.

Priestap (2023) outlines the dangers of sports betting.

> The fact that sports gambling is legal in so many states does not mean it is insulated from sophisticated criminals. In fact, the more money being wagered, the more unscrupulous people there will be seeking a piece of the pie. Such people will attempt to infiltrate (and influence) gambling companies. They will try to manipulate athletes competing in the games (to shave points), and they will do whatever it takes to obtain nonpublic information about individual players and teams, including bribery and blackmail as this insight can provide them an information advantage over others betting on the same contests.
>
> In pursuit of nonpublic information that will give them an edge, unscrupulous gamblers will do what organized crime groups and spy agencies have always done.

> They will identify people who have "access and a vulnerability," and then they will target and exploit some of them. Access in this context refers to people who possess nonpublic information about players and teams (which good player is more injured than reported). Vulnerability refers to those with a weakness of some type that could open them up to being taken advantage of by another person (a drug, gambling, or sex addiction, a major financial problem, etc.).
>
> In effect, an unscrupulous bettor will seek to satisfy the vulnerable person's desires in exchange for the access (i.e., nonpublic information) that the insider can provide. In his former line of work, Priestap regularly witnessed people being exploited—some wittingly, some unwittingly—by criminals and spies. These are tried and true tactics, and the same activity (will likely) become prevalent in the sports realm, all driven by gambling.

Violence by Players. McMurtry (1972) provides a compelling view of violence associated with American football.

> It is arguable that body shattering is the very point of football. In the United States, for example, the game results in fifteen to twenty deaths a year and about 50,000 major operations on knees alone. It is instructive to listen to the imperatives most frequently issued to the players by their coaches, teammates, and fans.

Their language includes "long bomb," "blitz," "take a shot," "front line," "pursuit," "good hit," and so on.

> Their principles and practices include mass hysteria, the art of intimidation, absolute command and total obedience, territorial aggression, censorship, inflated insignia and propaganda, blackboard maneuvers and strategies, drills, uniforms, formations, marching bands, and training camps. And the virtues they celebrate include hyper-aggressiveness, coolness under fire, and suicidal bravery.
>
> Of course, there is little or no protest against football. Perhaps the most extraordinary thing about the game is that the systematic infliction of injuries excites in people not concern, as would be the case if they were sustained at, say, a rock festival, but a collective rejoicing and euphoria. Players and fans alike revel in the spectacle of a combatant felled into semi consciousness, "blindsided," "closelined," or "decapitated." I can remember, in fact, being chided by a coach in pro ball for not injuriously hitting a player who was already lying helpless on the ground.
>
> After every game, of course, the papers are full of reports on the day's injuries, a sort of post-battle "body count," and the respective teams go to work with doctors and trainers, tape, whirlpool baths, cortisone, and morphine to patch and deaden the wounds before the next game. Then the whole drama is re-enacted—injured athletes held together by adhesive, braces, and drugs—and the days following it are filled with even more feverish activity to put on the show yet again at the end of the week. The team that survives this merry-go-round spectacle of skilled masochism with the fewest incapacitating injuries usually wins. It is a sort of victory by ordeal: "We hurt them more than they hurt us."

Violence by Spectators. Wen (2014) summarizes a sociological history of soccer violence. English soccer violence was prevalent from the 13th century on. Organized, rather than spontaneous, violence emerged from the 1960s to the 1980s. To counter such violence, physical restraints were added to separate spectators from players. Such restraints sometimes led to numerous deaths during riots – 39 in 1985 and 96 in 1989 at UK soccer matches. South American matches in 2013 led to 18, 40, and 30 deaths.

Wen concludes that

> Athletic events are realms in which other major issues in society, often related to class, religion, ethnicity, politics, regionalism, historic rivalries, etc. can play out among supporters. Violence, rather than just being about the sport, can be interpreted as an expression of contrasts between populations.

McKinley (2000) argues that a team is an extension of its fans.

> Ardent fans can become aggressive or even violent at a sports event when their team is threatened with losing. Fans sometimes throw batteries and other objects at players, or brawl with fans from the opposing team. A theory proposed to explain this behavior is that fans see the team as an extension of themselves, and so feel personally threatened by a loss.
>
> In Europe, where nationalist emotions often run high at soccer matches, violence is even more common. This sort of violence owes much to tribalism, psychologists say. Patriotism often mutates into overt violence at soccer matches, as exultant winners are humming with hormonal surges associated with victory and frustrated losers are looking for scapegoats.
>
> Some violence perpetrated by fans of a winning team may also be linked to surges in testosterone and feelings of invincibility fostered by soaring self-esteem, psychologists report. Indeed, an older theory – that watching violent sports serves as a catharsis that makes people less likely to be violent – has largely been discredited.

Summary. Gambling is a common behavior in our society. It cannot be eliminated, but we can try to eliminate the impacts of organized crime, both to avoid monetary exploitation of fans and to assure games and their outcomes can be trusted.

Violence is also a common phenomenon in our society. Sometimes it is inherent in the game, but rule changes and protective gear can lessen the consequences. Spectator violence is unacceptable, but perhaps predictable. Mitigating the consequences should be the goal.

Sociology of Online Games

Online games are increasingly pervasive and provide social environments for a steadily expanding population of players, driven by both the isolation resulting from the pandemic and steady technology advances. In this section, I first address the demographics of game players, ranging from university students to teenagers, to soldiers in Ukraine.

Discussion then addresses the social impacts of game playing, with emphasis on impacts more than who is playing. This includes consideration of potential negative impacts of game playing.

Withdrawal to only interacting in online worlds can result in "game disorder," and particular personality types seem prone to this malady.

Demographics of Game Players

Uz and Cagiltay (2015) report on university students' participation in online games and the resulting social environments that foster friendships and pre-existing relationships. They studied

> a sample of 168 university students to explore the social interactions in and out of the game environment in terms of personality type, gender and game preferences. It was found that participants mostly prefer playing multi-player games with their real-life friends and family members. While they tend to make friends in game environments, they do not prefer sharing sensitive issues with their gaming friends. Moreover, students who reported themselves as more extraverted, spend more hours in games.

Lenhart and colleagues (2015) present findings of an extensive survey of teen use of games, texting, phones, etc. Their key findings include:

- For today's teens, friendships can start digitally: 57% of teens have met a new friend online. Social media and online gameplay are the most common digital venues for meeting friends.
- Text messaging is a key component of day-to-day friend interactions: 55% of teens spend time every day texting with friends.
- Video games play a critical role in the development and maintenance of boys' friendships.
- Teen friendships are strengthened and challenged within social media environments.
- Some conflict teens experience is instigated online.
- Girls are more likely to unfriend, unfollow and block former friends.
- Teens spend time with their closest friends in a range of venues. Texting plays a crucial role in helping close friends stay in touch.
- Smartphone users have different practices for communicating with close friends, i.e., texting versus phone calls.
- Girls are more likely to use text messaging – while boys are more likely to use video games – as conduits for conversations with friends.
- Phone calls are less common early in a friendship, but are an important way that teens talk with their closest friends.

Gibbons-Neff (2023) reports on Ukrainian soldiers, currently in active combat assignments, playing *World of Tanks* to relax and interact with friends, as well as learn to work in a team and develop tactics in the game. "The urge to play a violent video game in the midst of the most brutal land war in Europe since World War II may seem baffling, but it represents an important way soldiers cope with the bloodshed around them: disassociation."

Social Impacts of Game Playing

Williams and colleagues (2018) present a comprehensive analysis of how role-playing games impact social organization, power structures, and team building.

> Beyond the experiences and interactions of players, role playing games are embedded in larger social and cultural processes and structures that enable and constrain play. Social processes refer to the organization of relations among games, designers, producers, players, and others, while cultural processes have to do with the organization of meanings, which both enables those relations and derives from them.
>
> Power structures gameplay, both formally and informally. Formal power structures include rules: Handbooks and rule books, dice, maps, and other materials support the designers' definitions of what constitutes "proper" play. Terms of service, player codes of conduct, and computer algorithms all structure gameplay. Roles, such as dungeon masters and guild leaders, formalize which members of the community have power, but power is also often facilitated informally through status hierarchies and norms within gaming groups. Younger or newer players, for example, may be sanctioned differently than more advanced players for breaking rules.
>
> Power is a practice in which players engage as they interpret and apply or negotiate rules and norms, make decisions on how to divide resources or loot, and treat players in various ways based on their perceived values or statuses. Individuals are given authority or leadership roles, but power can also be revoked or modified. Over time, players can gain or lose status either in-game or out-of-game; in fact, the two may be connected.
>
> Culture can be understood as an abstract and yet coherent web of meanings that people use to define themselves, their relations to each other, and the social actions that take place in and around games. Social player relations and cultures of play are intertwined: Many of the organizations and social groupings just mentioned also provide a sense of purpose and identity to players – very much a cultural phenomenon. Most gamers would agree that the identity of "being a gamer" involves specific interests and likes, caring about or valuing things in unique ways, and perhaps even talking and acting differently.
>
> In order for players to successfully get a game going, they must develop a shared understanding of what is going on (the situation), who they and the other players are (identities), and what kind of behavior is expected of them (role), all of which is guided by cultural knowledge and norms about different types of situations (frames), including layers of meaning (laminations), such as a player's speaking "in character" versus "out of character."

Lufkin (2020) explores the psychological and social benefits of gaming.

> With the rise of social media, gamers – particularly in Gen Z – have perfected the art of building communities in and around video games. Gamers don't just compete with strangers on the internet, but forge genuine, enduring friendships. (While) the concept of socialization in a game is new to many, video game enthusiasts have been using tech like this to build friendships online and stay connected for years. As mental health professionals stress the importance of relationships, connections and community in these times, they're even beginning to find direct psychological and social benefits from gaming across the generations.

Steinkuehler and Williams (2006) discuss how online games foster social engagement and a "third place" for informal sociability. They examine

> the form and function of massively multiplayer online games in terms of social engagement. Combining conclusions from media effects research informed by the communication effects literature with those from ethnographic research informed by a sociocultural perspective on cognition and learning, they present a shared theoretical framework for understanding (a) the extent to which such virtual worlds are structurally similar to "third places" for informal sociability, and (b) their potential function in terms of social capital.

They conclude that

> by providing spaces for social interaction and relationships beyond the workplace and home, online games have the capacity to function as one form of a new "third place" for informal sociability. Participation in such virtual "third places" appears particularly well suited to the formation of bridging social capital—social relationships that, while not usually providing deep emotional support, typically function to expose the individual to a diversity of worldviews.

Granic and colleagues (2014) summarize the positive cognitive, motivational, emotional, and social effects of playing video games – with friends, family, and complete strangers. They discuss the potential for interventions that promote well-being, including the prevention and treatment of mental health problems in youth.

> Video games today and those on the radar for development in the near future are also unique forms of play. Video games are socially interactive in a way never before afforded. Increasingly, players are gaming online, with friends, family, and complete strangers, crossing vast geographical distances and blurring not only cultural boundaries but also age and generation gaps, socioeconomic differences, and language barriers. The large amount of time invested in playing video games may also mean that they provide qualitatively different experiences than conventional games.

Potential Negative Impacts

Heng and colleagues (2021) discuss how games foster the creation of social capital, but can contribute to

> gaming disorder, an emerging mental health issue that should be further investigated. At present, there is a lack of agreement as to the precise name and definition of the online gaming disorder, which often is referred to as game addiction, pathological online game use, or problematic online game use. Gaming disorder means that excessive online gaming leads to gamers developing addiction-like symptoms (e.g., overuse) and negative consequences on physical/psychological health. This term describes the quintessence of the phenomenon (i.e., the behavior is not only excessive but fosters gaming-related problems).

Belongingness theory suggests that people have a fundamental need to belong that motivates them to seek out social interactions and form close and meaningful relationships with others. The social features of video games provide opportunities for new meaningful and emotionally resonant relationships to develop, helping to satisfy the human need for affiliation and social support. Therefore, social need and developing online relationships are main motivations for online gaming, and the social elements of an online game shape the gamers' desire to forge and maintain online relationships, which may play a considerable role in the initiation, development, and maintenance of gaming disorder. Consequently, the intensity of this social interaction has been known to be associated with gaming disorder.

High online social capital is indicative of a meaningful and emotionally supportive online community. However, social capital can also result in negative consequences. According to gratification theory, individuals' dependency on media is related to use gratification. Online social capital derived from in-game social interaction satisfies the need for affiliation and social support, which, in turn, leads to excessive gaming. Another mechanism through which online gaming may affect gaming disorder is suggested by the displacement hypothesis. Because of the "inelasticity of time", playing online games takes away time from face-to-face interactions with one's offline ties, which can lead to the displacement of offline social contacts for online ties.

Therefore, gamers who are absorbed with in-game social interaction may have an overall smaller and weaker offline social circle as a result of excessive online gaming. Reliance on online social interaction reduces offline contact, further maintaining online friendships and interactions. As gamers grow closer to their in-game contacts and their online social capital increases, offline activities become displaced and online game play becomes more desirable. Consequently, online gamers who participate in online social interaction might develop close ties with other gamers and receive social support from them, which, in turn, might lead to their psychological dependency on the online relationship, and the reduced levels of offline social interaction encourage the development of gaming disorder.

Research has found that problematic video game play was associated with significantly higher online social capital and lower offline social capital. Therefore, the benefits of in-game social interaction, namely, online social capital, can increase the risk for problematic behavior in the form of gaming disorder. Results showed that online social capital was a mediator in the relationship between in-game social interaction and gaming disorder. Moreover, for individuals with low alienation, the effect of online social capital on gaming disorder was weaker than for those with high alienation.

Summary

Game playing online is pervasive and growing rapidly among a wide range of populations, both younger and older. Most research reports substantial social benefits from playing online games. There is, however, the unfortunate possibility of addiction to online games and consequent withdrawal from face-to-face social interactions. The resulting social isolation is not healthy.

Table 5.1 Contrasts between Sports Games and Online Games

Attribute	*Sports Games*	*Online Games*
Well-Defined Objectives	Winning & money	Winning & fun
Well-Defined Rules	Regulated by league	Embedded in games
Dedicate Playing Spaces	Owned by each team	Hosted online
Professional Players	Must be drafted	Anyone can play
Minor League or Equivalent	Exists or relies on college	Not necessary
Published Annual Schedule	Required for TV schedule	Not useful
Championship Playoffs	Highly popular	Ad hoc, if at all
Broadcast Revenues	Enormous	Seldom relevant
Hall of Fame Elections	Highly popular	Ad hoc, if at all

Contrasting Perspectives

Table 5.1 contrasts sports games and online games. The entries in the sports games column reflect the organized business nature of sports game, even in college – but seldom in K–12. The online games column reflects a much more distributed and relaxed approach. There may be millions of players, but they are not all in the same instantiation at the same time. Yet, both columns reflect complex social systems with highly motivated and engaged players.

Conclusions

Politics is often included in the purview of sociology. I have avoided this topic in this chapter, instead addressing it in the next chapter. While this book is not concerned with political games, politics can certainly affect the games of interest in this book. Not only is the power of games pervasive, society broadly defined tends to have intimate interconnections with players, teams, franchises, leagues, etc.

References

Afshar, A. (2015). *Collegiate Athletes: The Conflict Between NCAA Amateurism and A Student Athlete's Right of Publicity.* Salem, OR: Willamette University, College of Law.

Blackstone, K.B. (2023). College football doesn't need realignment: It needs to start over. *Washington Post*, August 13.

Branch, T. (2011). The shame of college sports. *The Atlantic*, October.

Broch, T.B. (2022). The cultural sociology of sport: a study of sports for sociology? *American Journal of Cultural Sociology*, 10, 535–542. https://doi.org/10.1057/s41290-022-00177-y.

EDJNet (2020). Suspicious sports results? Mafia might have fixed them. *European Data Journalism Network*, November 18.

Edwards, H. (1969/2017). *The Revolt of the Black Athlete.* Urbana, IL: University of Illinois Press.

Gibbons-Neff, T. (2023). In a war of tanks, Ukrainian Soldiers play World of Tanks online. *The New York Times*, August 21.

Giulianotti, R., & Ansgar Thiel, A. (2023). New horizons in the sociology of sport. *Frontiers in Sports and Active Living*, 4, 1060622. https://doi.org/10.3389/fspor.2022.1060622

Godfrey, S. (2023). College football is barreling toward a super league, no matter what might be lost. *Washington Post*, August 15.

Granic, I., Lobel, A., & Engels, R.C.M.E. (2014). The benefits of playing video games. *American Psychologist*, 69 (1), 66–78.

Heng, S.P., Zhao H.F., & Wang, M.H. (2021) In-game social interaction and gaming disorder: A perspective from online social capital. *Frontiers of Psychiatry*, 11, 468115. https://doi.org/10.3389/fpsyt.2020.468115.

Lenhart, A., Smith, A., Anderson, M., Duggan, M., & Perrin, A. (2015). *Teens, Technology & Friendships*. Washington, DC: Pew Research Center. http://www.pewinternet.org/2015/08/06/teens-technology-and-friendships/.

Lufkin, B. (2020). How online gaming has become a social lifeline. *BBC Worklife*, December 16.

McKinley Jr., J.C. (2000). Sports psychology; A team is an extension of the fans. *The New York Times*, August 11.

McMurtry, J. (1972). Smash thy neighbor. *The Atlantic*, January.

Michener, J.A. (1976). *Sports in America*. New York: Dial Press.

Priestap, B. (2023). The dangers of sports betting. *Boston Globe*, February 14.

Root (2013). 'Schooled': Are college sports a modern-day plantation? *The Root*. https://www.theroot.com/are-college-sports-a-modern-day-plantation-1790898505.

Serra, P. (2015). The sociology of sport. *Principles of Sport Management* (Chap. 4). Oxford, UK: Oxford University Press.

Steinkuehler, C.A., & Williams, D. (2006). Where everybody knows your (screen) name: Online games as "Third Places". *Journal of Computer-Mediated Communication*, 11, 885–909.

UNODC (2023). *Global Report on Corruption in Sport: Illegal Betting and Sport*. Vienna: United Nations Office on Drugs and Crime.

Uz, C., & Cagiltay, K. (2015). Social interactions and games. *Digital Education Review*, 27, June.

Wen, T. (2014). A sociological history of soccer violence. *The Atlantic*, July 14.

Wikipedia (2023). Sociology of sport. *Wikipedia*. Accessed May 8, 2023.

Williams, P., Kirschner, D., Mizer, N., & Deterding, S. (2018). Sociology and role-playing games. In S. Deterding & J. Zagal (Eds.), *Role-Playing Game Studies* (Chap. 12). Abingdon, Oxfordshire, UK: Routledge.

Chapter 6

Politics of Games

Introduction

Games that cross political boundaries can precipitate political controversies. Domestic rivalries such as the Red Sox versus Yankees can be vituperative, but not on the scale of controversies that cross international boundaries. This chapter addresses the political implications of three sports – the Olympics, World Cup soccer, and Chess. These sports seem to deeply engender national pride or possibly melancholy, depending not only on who wins, but also who is judged respectful or not.

All three sports provide ample evidence of "sportswashing" – the practice of individuals, groups, corporations, or governments using sports to improve reputations tarnished by wrongdoing. A form of propaganda, sportswashing can be accomplished through hosting sporting events, purchasing, or sponsoring sporting teams, or participating in a sport. I elaborate this phenomenon for each of the three sports.

Olympics

In April of 1896, the Olympic Games, a long-lost tradition of ancient Greece, were reborn in Athens, 1,500 years after being banned by the Roman Emperor. French baron, Pierre de Coubertin, had proposed reviving the Olympics as a major international competition that would occur every four years. At a conference on international sport in Paris in June 1894, Coubertin again raised the idea and it was unanimously approved. The International Olympic Committee (IOC) was formed, and the first Games were planned for 1896 in Athens, Greece.

Boykoff (2016) provides an in-depth and insightful history of the Olympics.

> In reality the Olympics are political through and through. The marching, the flags, the national anthems, the alliances with corporate sponsors, the labor exploitation behind the athletic-apparel labels, the treatment of indigenous peoples, the marginalization of the poor and working class, the selection of Olympic host cities—all political. To say the Olympics transcend politics is to conjure fantasy.

DOI: 10.4324/9781003491927-6

Critical engagement with the politics of sports has historically helped pry open space for ethical commitment and principled action, as evidenced by the Olympic athletes who have taken courageous political stands, the alternatives to the Olympics that have emerged over the years, and the activism that springs up today to challenge the five-ring juggernaut. (However), the kind of passion sports generate can be channeled in countless directions, from the radical to the reactionary, from reverence to treachery.

The Olympic Games have become a cash cow that the IOC and its corporate partners milk feverishly every two years, since the staggering of the Summer and Winter Olympics began in 1994. For the IOC, acknowledging politics might jeopardize their lucre. The International Olympic Committee is a well-oiled machine, with slick PR, palatial accommodations in Lausanne, Switzerland, and around $1 billion in reserves.

Timeline of Games

At the turn of the century the Olympics did not yet enjoy the cachet they have today, so the 1900 Games in Paris and the 1904 Games in St. Louis had to affix themselves to the enormous cosmopolitan institutions of the day—World's Fairs. The early modern Olympics were mere sideshows to the World's Fairs, not the main event on the world stage that we see today.

Many historians identify the 1912 Games in Stockholm as the Olympics that established them as a top-tier international event. Jim Thorpe was a dazzling multi-sport athlete who starred in football, baseball, and track and field. In 1912 he achieved the remarkable feat of winning both the pentathlon and the decathlon. One year after the Games, Thorpe was stripped of his medals for having broken the amateur code—in 1909 and 1910 he received a small sum of money ($60 a month) for playing semiprofessional baseball.

This 110-year-old injustice was recently corrected and his gold medals restored for the pentathlon and decathlon (Wu, 2022).

World War I forced the cancellation of the Sixth Olympiad, which had been scheduled to be in Berlin in 1916. The Olympics returned after the Great War, to Antwerp in 1920. Paris hosted the 1924 Olympics, Amsterdam the 1928 games, and Los Angeles the 1932 games. The infamous Berlin games were in 1936.

The Olympics were initially of little interest to Adolf Hitler. When the 1936 Games were awarded to Berlin in 1931, a centrist, democratic coalition held power in Germany. Even in 1932, Hitler was referring to the modern Olympics as "a plot against the Aryan race by Freemasons and Jews."

(Boykoff, 2016)

Hitler loomed large in the Olympic stadium, but US track star Jesse Owens ruled the athletics oval, winning four gold medals. Owens, the son of an African-American Alabama sharecropper who moved his family north to Cleveland, Ohio, in search of opportunity, was the indisputable superstar of the Berlin Games.

Owens had emerged as a track and field sensation in the States. He tied the world record in the 100-yard dash while still in high school, and his performance at the 1935 Big Ten Championships, in which he established three world records and matched a fourth over a span of 45 minutes, remains one of the most extraordinary accomplishments in collegiate sports history (Ott, 2021).

After setting an Olympic record in the 200-meter dash en route to a third gold medal, Owens put the exclamation point on his showing by running the opening leg of a record-shattering US 4×100 relay performance. He became the first American of any race to win four gold medals in track and field in a single Olympics, an achievement that stood unaccompanied until Carl Lewis matched him in 1984.

There was an attempted boycott of these Olympics by the US, Great Britain, France, Sweden, Czechoslovakia, and The Netherlands. There was strong support for the boycott in the US, but a vote of the Amateur Athletic Union in the US failed to support it. A People's Olympiad was planned for Barcelona, Spain but was canceled with the onset of the Spanish Civil War.

The 1936 Berlin Games were laced with controversy.

> This was due to the fact that the Nazis had a hatred for anyone who was not an Aryan, people with blonde hair, blue eyes, athletic and tall. Many people were surprised that people of other races besides Aryans were able to participate in the Berlin games.

Moreover, the "Nazis were deeply offended by sporting contacts with 'primitive' races and by competing against Negro athletes, in particular."

Adolf Hitler would go on to voice how he was in agreement with the concept of segregation for interracial athletic competitions, because the people who had ancestors that "came from the jungle were seen as primitive because their physiques were stronger than those of civilized whites." Since the Nazis were unable to segregate the races, they used their hatred for non-Aryans as fuel that allowed them to use the 1936 Berlin Games as a way to assert dominance against the inferior races (USHMM, 2023).

The 1940 Olympics were scheduled for Tokyo, moved to Helsinki, then canceled, as were the 1944 Olympics that were tentatively scheduled for London, but delayed until 1948. The 1952 Olympics were held in Helsinki. With the emergence of the Cold War, the Olympics became a prominent venue for rival political systems to assert their superiority.

The 1956 games in Melbourne were boycotted by Egypt, Iraq, and Lebanon for the British and French incursion of the Suez Canal, by The Netherlands and Spain for Soviet suppression of Hungarians, and by China for recognition of Taiwan. For the 1960 games in Rome, Taiwan was represented as Taiwan, not China. Consequently, Taiwan withdrew from the games. China established a competing Games of the New Emerging Forces.

South Africa's apartheid system led the IOC to withdraw the country's invitation to the 1964 Tokyo Games and the 1968 games in Mexico City. The IOC ultimately expelled South Africa from the Olympic Movement in 1970, only to reinstate it in 1992.

Harry Edwards published *The Revolt of the Black Athlete* in 1969. It chronicles the Olympic Project for Human Rights, which facilitated the Black Power Salute protest by two African-American athletes, Tommie Smith and John Carlos at the 1968 Summer Olympics in Mexico City. This book was published again in 2017 with an extensive new introduction (Edwards, 2017).

The 1972 games in Munich included the Black September mass killings of Israeli athletes. Eleven members of Israel's Olympic team and four Arab terrorists were killed in a 23-hour drama that began with an invasion of the Olympic Village by Arabs. It ended in a shootout at a military airport some 15 miles away as the Arabs were preparing to fly to Cairo with their Israeli hostages (Binder, 1972).

After the terrorists killed two Israelis and took control of the Olympic Village, a number of events continued until activities were suspended late in the day. Despite the carnage at the airport, Avery Brundage, chairman of the International Olympic Committee, famously said, "The Games must go on," and 34 hours later competition resumed.

The Montreal Games in 1976 ended up costing US$1.5 billion, making them the most expensive Olympics to that point, and saddling Canadians with a debt that would take them three decades to pay. These 1976 Games also brought a leap in Olympic sponsorship. When the federal government refused to cover the Olympics' fiscal deficit, Montreal organizers went full-on corporate, teaming with commercial "sponsors" and "official suppliers.. The business side of the games was now fully energized.

The US and 54 other nations boycotted the 1980 games in Moscow because of the Soviet invasion of Afghanistan. US President Jimmy Carter said he was not "naïve," and that he realized that the alternative games the Administration is attempting to organize would be no substitute to Olympic competition. But he urged the athletes to regard the boycott positively as a means of

> having helped to preserve freedom and having helped to enhance the quality of the principles of the Olympics, and having helped in a personal way to carry out the principles and ideals of our nation, and having made a sacrifice in doing it.
>
> **(Scannell, 1980)**

The Soviets and Eastern bloc countries boycotted the 1984 games in Los Angeles in response to the US boycott in 1980. The Soviet Union announced that it would not take part in the 1984 Summer Olympic Games in Los Angeles because the Reagan administration "does not intend to ensure the security of Soviet athletes." A statement by the Soviet National Olympic Committee accused the Reagan administration of being in "direct connivance" with various extremist organizations seeking to create "unbearable conditions" for Soviet participants (Doder, 1984).

The Los Angeles games was the first time in Olympic history that a private group, rather than the city itself, would organize the Games. Peter Ueberroth led this. Ueberroth's corporate organizing committee took responsibility for cost overruns. Members of the committee believed that the privatized nature of the Games "made them a powerful instrument to demonstrate the validity of the American free enterprise system."

Nevertheless, the advocates of privatization quietly reaped public subsidies. Public services arrived gratis, as did an enormous flock of unpaid volunteers who did much of the grunt work. The federal government contributed to security for the Games. Transportation networks were already extant, as was the publicly funded communications structure. In the end, the Los Angeles Olympics reaped a profit of about US$215 million.

"The tight relationship between big business and the Olympics foreshadowed what has been described as a multi-billion dollar exercise in global marketing locked up and ring-fenced by saber-toothed attorneys" (Boykoff, 2016).

The 1988 games were in Calgary and the 1992 games were in Barcelona. In Barcelona, "The games were put at the service of a preexisting (urban development) plan, rather than the typical pattern of the city development plan being put at the service of the games." Costs ballooned from US$600 million to US$11 billion.

For the 1996 games in Atlanta,

> ACOG financed the Olympics $2.2 billion and walked off with a small financial surplus, though this ignores the approximately $2 billion spent by public authorities including $996 million in federal government investment, $226 million in state funds, and $857 million in local funds. Tragedy struck the Games when a bomb blast ripped through Centennial Olympic Park, killing one person and wounding more than one hundred others. Atlanta also provided a foretaste of the feast of corruption to come.
>
> Conservationism made its splashy debut at the 2000 Sydney Games. Organizers insisted that environmental progress "was one of the shining achievements" of the Olympics, "a hallmark of Sydney's Games. Social sustainability" in all its nebulousness was another buzzword at Sydney. The Sydney Games may have evoked feelings of goodwill and reconciliation, but they left a $1.7 billion debt in their wake, an unwanted legacy for the taxpaying public.
>
> **(Boykoff, 2016)**

The 2002 Winter Olympics in Salt Lake City exposed a long suspected bias in judging ice skating events. A French judge was confronted about her judging of the pairs free skating event. She broke down in a tearful outburst that was witnessed by a number of other skating officials who happened to be present. She said that she had been pressured by the head of the French federation to put the Russians first as part of a deal to give the ice dancing gold to the French ice dance team. She has never judged a competition since. This is a risk for any sport that lacks objective measure of success.

The 2004 games in Athens ended up costing Greece US$16 billion. The bulk of budget-busting expenditures included venue construction, infrastructure development, and security. The 2008 games in Beijing included a surreptitious agenda. "Behind the convenient smokescreen of terrorism prevention, the Chinese government spent billions to retool its repressive apparatus. The five-ring price tag was more than $40 billion. Tourism actually declined in China during the year of the Games."

The 2012 games were in London and the 2016 games in Rio. The World Cup (in 2014) was a foretaste of what was to come with the Olympics. "The Olympics are a much more complicated endeavor than the World Cup. Rio 2016 featured more than 10,000 athletes from 205 countries, whereas the Brazil World Cup involved 736 soccer players from thirty-two nations" (Boykoff, 2016).

> For the first time in sixty years, the two Koreas stood undivided. The Pyeongchang Olympics (Winter Games in 2018) will be marked in history as a symbolic breakthrough for supposedly all Koreans bursting in patriotism. We even glorified it to the extent of shifting South Korea's policies toward imminent reunification. But did one unified women's hockey team and flag really alleviate the tensions in the Korean Peninsula?
>
> It is patently obvious that the Winter Games in Pyeongchang were used merely as a "public image makeover" to gain political leverage while the status quo remains greatly unchanged. North Korea is not interested in diplomatically giving up its

nuclear program or reunifying unless the conditions are favorable to the North Korean regime.

(Yang, 2018)

One may think of an Olympic boycott as countries staying home, athletes and all. But the U.S. diplomatic boycott of the Beijing Winter Olympics in 2022 precluded only government officials from attending. In announcing the decision the White House cited "genocide and crimes against humanity" in Xinjiang, a northwestern region of China. The Chinese government has cracked down harshly on Uyghurs and other predominantly Muslim ethnic minorities in that region, including mass detentions and forced use of contraception and sterilizations. Calls for an Olympic boycott sharpened after Peng Shuai, a Chinese tennis star, accused a former top government official of sexually assaulting her.

(Mather, 2022)

Overall Observations

Boykoff (2016) offers a wealth of insightful overall observations about the Olympic enterprise.

Today the Olympic Games are an enormous sports, media, and marketing juggernaut, a top-tier athletic festival awash in corporate cash. The IOC sits at the center, oscillating Janus-faced between multinational conglomerate and global institution.

If modern sport is "capitalism at play," then the Olympic Games are particularly instructive for mapping the terrain of capitalism. An economic juggernaut, the Olympics do not merely symbolize or reflect capitalism; they actively produce it

Celebration capitalism is a political-economic formation marked by lopsided public–private partnerships that favor private entities while dumping risk on the taxpayer. A trend under celebration capitalism: the public bails out private entities when they flop under pressure.

The Olympics are an elite-driven affair with scant opportunity for meaningful public participation. Olympic elites insulate themselves from civil society. Celebration capitalism deploys state actors as strategic partners, putting forth public–private partnerships—rather than full-on privatization—as the dominant mode of economic transaction.

Academic sports economists have come to the general conclusion that hosting the Olympics does not generate the benefits that economic impact studies promise. The games lose large amounts of public money and add to public sector debt.

Fewer and fewer cities are open to the Games. For too long host cities have worked in service of the Olympics. It's time for the Olympics to start working in service of host cities. However, the price of hosting the Games is also sky-high because Olympic development is often disconnected from city and regional planning.

Four key risks to the Games: terrorism, protest, organized crime, and natural disasters. The role of security officials is to prevent this mishmash of "key risks" from marring the celebration.

World Cup

The first FIFA World Cup was held in 1930 in Uruguay, which defeated its nemesis Argentina to gain the championship. The next World Cup was the first of several highly controversial World Cups. It was hosted by Italy in 1934, which won the championship. This World Cup happened in a different world.

> There was no Amnesty International. There was no Universal Declaration of Human Rights. There had not yet been a 1936 Olympics, hosted in Nazi Germany.
>
> This is why Benito Mussolini's World Cup also opened football to a new world, and can almost claim to be the first of these types of sportswashing events. It was why the depth of political pageantry was much more overt, to the point it could almost be called pantomime as Mussolini had long realized the emotional and propaganda power of football and the World Cup, and how it also fitted with his classically fascist ideal of the Italian man as athletically virile.
>
> **(Delaney, 2022)**

The next virulent instance of sportswashing was in Argentina in 1978.

> Within view of the Estadio Monumental pitch there was the Navy Mechanical School that was at that point being used as a torture prison by Jorge Videla's regime. Of the 5,000 prisoners held there between 1976 and 1983, only 150 survived. Many were raped, castrated, and set upon with dogs and electric batons. All of this continued during the World Cup, where the prisoners could often hear the roar of the crowd at Estadio Monumental, just 500 meters away. They were known as "the disappeared", and it did mean that football could not completely insulate itself from the brutal reality of the host regime.
>
> **(Delaney, 2022)**

In May 2015, 14 people were indicted in connection with an investigation by the United States Federal Bureau of Investigation and the Internal Revenue Service Criminal Investigation Division into wire fraud, racketeering, and money laundering, including US$150 million in bribes. The United States Attorney General simultaneously announced the unsealing of the indictments and the prior guilty pleas by four football executives and two corporations. Nine FIFA officials and five corporate executives were indicted for racketeering conspiracy and corruption.

The choice of Russia as host of the 2018 World Cup – the third controversial selection – was challenged by many. Central issues included the level of racism in Russian football and the discrimination against LGBT people in wider Russian society. Russia's involvement in the ongoing conflict in Ukraine also caused calls for the tournament to be moved, particularly following the annexation of Crimea and support of separatists in the war in Donbas. Then FIFA President Sepp Blatter declined requests for the tournament to be moved.

The fourth controversial hosting was Qatar in 2020. Panja and Smith (2022) outline the controversy.

> American investigators and FIFA itself have since said multiple FIFA board members accepted bribes to swing the vote to Qatar. A broad corruption investigation into how

> FIFA conducts business led to dozens of arrests. Those cases and others helped bring down the entire leadership of FIFA, and almost toppled the institution itself.
>
> More accusations of corruption and bribery followed. The United States Department of Justice accused three South American voters of accepting seven-figure bribes to select Qatar. Within a few years, in fact, almost every one of the 22 members of the committee who had participated in the vote had been accused of or charged with corruption. Dozens of other executives had been arrested. Most were forced out of FIFA, and several were barred from soccer altogether.
>
> Even those at the very top of the rotten pyramid had not escaped. Blatter grudgingly announced he would resign, then was banned anyway. Platini was forced out, too, over an unrelated ethics charge that led to a fraud trial in Switzerland. (He and Blatter were both acquitted.) For a while, it seemed as if FIFA itself might not survive a decision of its own making.
>
> It (was) too late for that. As the tournament neared, the criticism of FIFA's decision to take it to Qatar only grew more pointed. An expanding list of current players, former players, coaches, sportswear manufacturers and, in particular, fans have been vocal in their opposition. The captains of England and Wales have agreed to wear a special armband promoting gay rights. Blatter, as recently as the month before the games, admitted the choice of Qatar was a "mistake."
>
> Qatar's response, in turn, (was) become steadily more indignant. The country's emir, the crown prince, lashed out at what he described as an "unprecedented" campaign of criticism from the West. Qatar's foreign minister labeled the questions over its suitability to host the tournament "very racist."

FIFA has not always been so opposed to the idea of using soccer for ideological purposes. Even after all of the investigations, the warrants and the arrests, FIFA as an institution has always justified its decision to go to Qatar by insisting that the sport can be an agent for progress.

Knight (2022) reports on Qatar's investment decision in detail.

> The Qatari edition was born in corruption, paid for with hydrocarbons, and built on the labor of hundreds of thousands of workers, imported from the Global South and frequently abused in one of the smallest and richest countries on earth.
>
> FIFA awarded Qatar the rights to host the World Cup on December 2, 2010. On the same day, the organization's executive committee voted to give Russia the 2018 edition. Of the twenty-two men who voted, fifteen were later indicted by American or Swiss prosecutors, banned from soccer, charged by FIFA's ethics committee, or expelled from the International Olympic Committee.
>
> The Qatari Investment Authority, which manages an estimated four hundred and fifty billion dollars, didn't build a stage for a soccer tournament; it built a city to encompass the stage. The World Cup cost more than two hundred billion dollars (that's around sixty times the expense of the 2010 tournament, in South Africa), but the price tag included the metro system, an airport extension, bridges, man-made islands, fighter jets, a collapsible stadium, and a bulk order of five-star hotels.

> Qataris make up about twelve per cent of the country's population—a ruling class of around three hundred thousand people. Of the 805,810 workers in the construction sector in 2017, 0.0016 per cent were Qatari nationals.
>
> People also talk about "sportswashing," to describe the activities of sovereign wealth funds like Saudi Arabia's, which recently acquired Newcastle United, an English Premier League soccer team, and plans to spend two billion dollars on a breakaway golf league. The term suggests using sports to launder a lousy reputation. But, in the case of Qatar, staging the World Cup was more about gaining a reputation at all.

"What is relevant as regards Qatar – this fourth overly political World Cup – is how visible all this was; how the competitions were used; how the players felt; how they coped and how much any achievement was sullied" (Delaney, 2022). Soccer was only one component of the overall endeavor.

Chess

There are 605 million regular chess players globally. "Chess has long been a game rife with allegations of chicanery and skullduggery. Cheating at chess is as old as the game itself" (Taylor, 2022).

Russia has played a central role in much of the chess drama (LePrince-Ringuet, 2018). Lenin's love of the game contributed to it becoming the Russian national pastime. Between 1948 and 1993, Russia produced seven of eight world champions. Bobby Fischer was the only non-Russian champion.

> U.S. prodigy Bobby Fischer accused Soviet players of deliberately drawing their games to preserve their energy for games against him. Four decades later, the former head of the Soviet team that year admitted the allegations were true.
>
> **(Taylor, 2022)**

> Fischer would go on to win the 1972 World Chess Championship, beating the Soviet player Boris Spassky. The U.S. victory was short-lived: Fischer would refuse to defend his title in 1975 and entered a long period of decline. In 1992, he won an unofficial rematch against Spassky but ended up facing an arrest warrant for breaking U.N. sanctions on Yugoslavia, where the match was held.
>
> In the 1993 World Open in New York, an unrated player who was able to force a draw against a grandmaster was accused of using technology to cheat. The player reportedly wore headphones, had a pulsing bulge in his pocket and appeared to not fully understand the basic rules of chess.
>
> Since then, the risk of technological cheating has affected chess at all levels. Three top French players were suspended for allegedly cheating via coded text message in 2011. Four years later, a Georgian champion was found to have an iPhone hidden in a bathroom during the 17th annual Dubai Open Chess Tournament.
>
> In modern chess, even the best players are no match for chess programs that can run on a phone. Garry Kasparov, the legendary Russian player, was able to beat IBM's

> supercomputer Deep Blue in 1996, but he became the first world champion to lose a match to a computer the following year, when Deep Blue won a rematch.

McClain (2014) outlines the intrigues and conspiracies of the Russian chess ecosystem. Corruption is pervasive across national chess federations, but in Russia the highest level politicians, e.g., President Vladimir Putin, are fully involved in the intrigues.

> At stake is not the leadership of some powerful country but the presidency of a fairly obscure organization that presides over a small corner of the gaming world, the World Chess Federation (FIDE). The body oversees international chess championships and controls tournaments and sponsorship deals worth millions of dollars and championships that are the grail of nationalistic aspirations.
>
> **(McClain, 2014)**
>
> The principal characters also seem drawn from fiction. There is a former world chess champion, now a Russian opposition leader; a former president of an obscure Russian republic who believes that he was abducted by extraterrestrials in yellow suits who invented the game of chess; and an ex-fashion photographer turned chess official who would like the first two candidates to be disqualified so that he can take over the federation.
>
> The latest intrigue revolves around corruption allegations by the two candidates for the federation's presidency, Garry Kasparov, the former champion and Russian opposition figure Kirsan Ilyumzhinov, the incumbent president and self-described space-alien abductee.
>
> The chess federation is no stranger to intrigue or to machinations by former players with towering egos and a thirst for strategic conquest. The key players all have strong ties to Russia, where chess is a matter of national pride and powerful political interest.
>
> While there has been no sign of Kremlin involvement in the recent elections, the Russian government has stepped in before on behalf of Mr. Ilyumzhinov, who was appointed by the Kremlin as president of the Russian republic of Kalmykia, and against Mr. Kasparov, a noted Kremlin political opponent.
>
> In 2010, the former world champion Anatoly Karpov ran for the presidency of FIDE with the endorsement of the Russian Chess Federation and Mr. Kasparov's backing. But after a top adviser to then-president Dmitri A. Medvedev sent armed guards to seize control of the Russian Federation's offices, the federation switched its endorsement to Mr. Ilyumzhinov, who went on to win re-election.

More recently (LePrince-Ringuet, 2018), Kirsan Ilyumzhinov was suspended in July 2018 for ethics violations, after serving seven terms for 23 years as President of FIDE. During this period, he assured his re-election by buying votes of Federation members. The violation that led to his suspension involved creating a fake (non-existent) Vice President candidate as his running mate for the upcoming elections.

Arkady Vladimirovich Dvorkovich, an economist, was elected President of FIDE in 2018. He was re-elected for a second term in the elections held at the 44th Chess Olympiad held in India in 2022. No further controversies have emerged of late.

Table 6.1 Prevalence of Corruption in Olympics, World Cup, and Chess

Corruption	*Olympics*	*World Cup*	*Chess*
Sportswashing	√	√	√
Election of Governing Boards	√	√	√
Selection of Tournament Sites	√	√	√
Play of Games			√
Judging of Games	√		

Comparison

Table 6.1 summarizes the assessment of corruption in the Olympics, World Cup, and Chess ecosystems. As noted earlier, sportswashing is pervasive as countries seek credibility despite unfortunate practices, such as human rights violations. Election of governing bodies and selection of tournament sites are subject to collusion and bribery as various stakeholders seek a share of the enormous amounts of money involved, especially in the Olympics and World Cup.

Cheating by game players is reported more often in chess than in the other two game ecosystems. Collusion in judging games has been more common in the Olympics for the sports judged by subjective ratings rather than objective measurements, for example, ice skating as I mentioned earlier.

Conclusions

These three sports represent grand events, particularly the Olympics and World Cup. Politically, the opportunities for sportswashing are pervasive. Governments hope that impressive and compelling games will distract people from government and societal shortcomings and failures in other arenas.

Financially, enormous sums of money are involved, attracting all sorts of schemes. NBC contracted with the IOC for US$7.75 billion for 20 years, thus ten Summer and Winter Olympics, averaging US$775 million per game. Fox and Telemundo contracted with FIFA for US$200 million per year. As I mentioned earlier, the Olympics is a much larger endeavor than the World Cup.

Not surprisingly, everybody wants a piece of the revenue pie. Bribes to be elected to governing bodies and to be selected as the games site provide other cash flows. Opportunities for corruption and bribery are central to the politics of games. Of course, these phenomena have long played out in society.

References

Binder, D. (1972). The Munich massacre. *The New York Times*, September 5.
Boykoff, J. (2016). *Power Games: A Political History of the Olympics*. London: Verso.
Delaney, M. (2022). A political World Cup is nothing new. *The Independent*, November 14.
Doder, D. (1984). Soviets withdraw from Los Angeles Olympics. *Washington Post*, May 9.
Edwards, H. (2017). *The Revolt of the Black Athlete*. Urbana, IL: University of Illinois Press.

Knight, S. (2022). At Qatar's World Cup, where politics and pleasure collide. *The New Yorker*, December 3.
LePrince-Ringuet, D. (2018). How chess became a pawn in Russia's political war games. *WIRED UK*, November 20.
Mather, V. (2022). The diplomatic boycott of the Beijing Winter Olympics, explained. *The New York Times*, February 6.
McClain, D.L. (2014). Intrigue, conspiracy, alien abduction: Politics of chess go off the board. *The New York Times*, February 8.
Ott, T. (2021). How Jesse Owens foiled Hitler's plans for the 1936 Olympics. *History*, June 10.
Panja, T., & Smith, R. (2022). The World Cup that changed everything. *The New York Times*, November 19.
Scannell, N. (1980). Carter tells athletes decision on Olympics is final. *Washington Post*, March 22.
Taylor, A. (2022). In chess, a long history of cheating, chicanery and Cold War shenanigans. *Washington Post*, October 5.
USHMM (2023). The Nazi Olympics. *Holocaust Encyclopedia*. https://encyclopedia.ushmm.org/content/en/article/the-nazi-olympics-berlin-1936. Accessed August 25, 2023.
Wu, T. (2022). Jim Thorpe's 1912 Olympic gold medal are finally reinstated. *Smithsonian Magazine*, July 20.
Yang, J. (2018). Politics and the Olympics. *The New York Times*, June 4.

Chapter 7

Games for Learning

Introduction

Games can play a central role in educating everybody. Indeed, younger people now spend more time playing games than watching television. This chapter addresses the education of three populations: K–12 students, college students, and employees. Each population has three subpopulations, yielding a total of nine targeted populations. Learning objectives for each of these populations are discussed. Alternative game-based interventions for achieving these objectives are considered.

Table 7.1 categorizes the nine targeted populations, as well as example learning objectives. This chapter considers alternative game-based interventions for achieving these objectives.

K–12 Students

We recently completed a two-year study of alternative policies to enhance the STEM (Science, Technology, Engineering & Mathematics) talent pipeline. We first addressed retention in STEM college programs, currently about 50%, and explored student support interventions that would likely yield better outcomes. We then addressed K–12 and how the percentage of graduates who are "STEM ready" might be improved from its current 16%.

We wondered what we could do to help students and teachers that would not require permissions from 14,000 local school boards, administrators of 100,000 high schools, and two large teacher unions. Figure 7.1 summarizes the interventions, often technology-enabled, that we chose to explore.

Table 7.2 summarizes the findings from a search on (internships OR camps OR games) AND (K–12 AND STEM AND students). The search and analysis processes resulted in 26 highly qualified articles. Table 7.2 summarizes the ten articles on games.

Overall, we found that camps, games, and projects that provide well-designed, engaging student experiences provide moderate positive impacts on knowledge and skills gained, personal ability benefits, and personal attitude benefits. These findings are limited to outcomes occurring during the intervention or immediately after, rather than long-term impacts months or years after the interventions, such as were discussed for training simulators in Chapter 2.

DOI: 10.4324/9781003491927-7

Table 7.1 Target Populations for Educational Games

Population	*Example Learning Objectives*
Elementary School Students	Counting, arithmetic, reading
Middle School Students	Geography, history, science, algebra
High School Students	Physics, chemistry, biology, calculus
All Undergraduate Majors	Real-world phenomena
STEM Undergraduate Majors	Work tasks & practices
STEM Graduate Students	Professional practices
Entry-Level Employees	Domain-specific work practices
Mid-Level Employees	Real-world phenomena
Senior Employees	Elements of competitive strategy

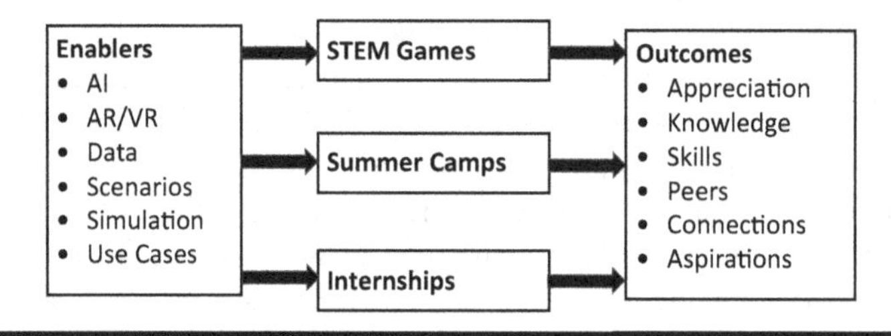

Figure 7.1 Immersive STEM experiences.

Figure 7.2 summarizes the relationships among the overall outcomes (Rouse, Lombardi, & Gargano, 2023). Attitude in terms of motivation, engagement, and self-perceptions is the key to fostering abilities and knowledge and skills. Interventions need to be sufficiently compelling to capture students' interests in ways that they feel empowered to act.

As I briefly noted in Chapter 4, engagement can be sustained for durations of a week or so and have greater impacts than longer interventions that lose their novelty. However, games with multiple levels can sustain interest if carefully designed and evaluated. Successive levels need to unveil greater challenges, more information, and additional tools.

For the games reviewed, there were limited gender differences as the interventions were designed with such possibilities in mind. It appears that gender differences emerge in environments where the designers were not sensitive to these phenomena, e.g., by designing all the characters as males.

Figure 7.3 summarizes the types of games reported in the literature reviewed. Design environments involve students in designing and evaluating solutions. Examples include coding games, immersive virtual reality (VR) for biological gene transcription, and planning for interactive liquid-handling robots. Example interactive games include *River Crossing*, *Treasure Hunt*, and *Tower of Hanoi*. Instances of immersive games include immersive VR for flipped classrooms.

Board games and video games tend to be targeted at younger children, often involving familiar games and video games that conclude with an online quiz. A few games involve interactive models and simulations, for example, science concept acquisition. For many games, the models and

Table 7.2 Summary of Findings

Article	*Intervention*	*Outcomes*
Agbo et al. (2022)	VR Games – *River Crossing, Treasure Hunt, Tower of Hanoi*	VR increased students' motivation and computational thinking skills, via immersion, interaction, and engagement in the VR educational application, leading to higher cognitive benefits, increased interest, and attitude to learning computational thinking concepts.
Al-Zaytoonah (2016)	First Grade Science Students – Science Concept Acquisition	Eight educational games were developed, and a test to measure scientific concepts acquisition. Results showed that there were statistically significant differences in students' scientific concepts acquisition due to the method of teaching in favor of the group using the games. There were no statistically significant differences in students' scientific concepts acquisition due to the gender.
Aleman et al. (2022)	University Makerspace Games	Students carry forms of capital that impact their entry into learning spaces as they accumulate capital through everyday talk and storytelling. Findings point to the critical role of intentional communication and space design in cultivating inclusive makerspace cultures
Artzmann et al. (2022)	Literature Review – 39 Studies of Science, Math & Engineering	A meta-analysis across 39 independent studies compared game-based learning interventions with traditional classrooms in primary and early secondary STEM education. Moderate positive effects on cognition, motivation, and behavior were found. Primary school students achieve higher learning outcomes and experience game interventions as more motivating than secondary school students, whereas gender did not have any moderating effect.
Bertram (2020)	Math & Computer Science Skills	Core elements of psychological theories of learning are presented, including arguments for and against digital learning games. This leads to suggestions for game content, which would be suitable for preparing students for university-level mathematics and computer science education, and discuss the potential limitations of digital learning in the classroom.
Elme et al. (2022)	Immersive VR for Biology – Gene Transcription	The IVR simulation led to a significant increase in knowledge from the pre- to post-test, but there were no differences between the self-explanation and control groups on knowledge gain, procedural, or conceptual transfer. The self-explanation group reported significantly higher intrinsic cognitive load and extraneous cognitive load.

(*Continued*)

Table 7.2 (Continued) Summary of Findings

Article	*Intervention*	*Outcomes*
Jong (2022)	Immersive VR for Flipped Classroom	Results indicated the participants across the three majors positively perceived spherical video-based immersive virtual reality as having desirable benefits on attention, relevance, and satisfaction, but not confidence.
Lil et al. (2022)	Interactive Liquid Handling Robots	Affordable and accessible liquid handling robots provide a useful educational tool to be deployed in classrooms, and robot-based curricula may encourage interest in STEM and effectively introduce automation technology to life science enthusiasts.
Wang et al. (2022)	Literature Review – 33 Studies of Science, Math & Engineering	Meta-analysis results across 33 independent studies demonstrated that compared to non-digital game learning activities, digital game-based learning had a moderately significant effect on students' STEM learning achievement. Using digital games to improve students' academic performance could be one of the effective methods for STEM education. Intervention durations of less than one week have the largest effect size.
Zhang et al. (2022)	Coding Games	Coding motivation and feeling of enjoyment were predictors of the actual use of the game, with coding motivation the dominant factor. Qualitative findings supported the quantitative results – students who were more intrinsically motivated tended to be more active in using the game.

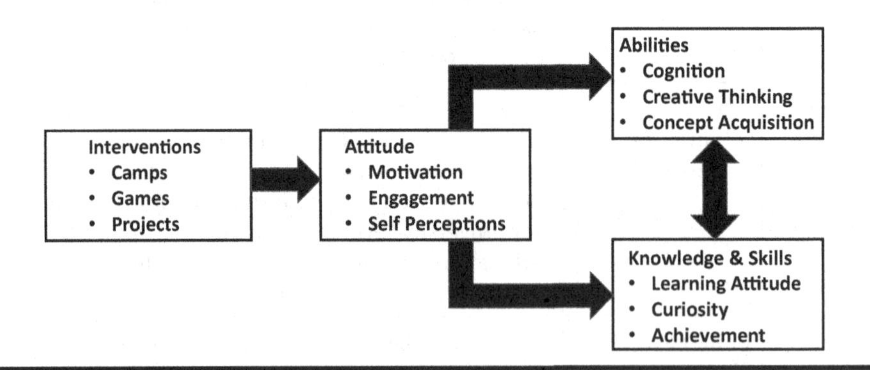

Figure 7.2 Relationships among interventions, attitude, and abilities.

simulations are embedded in the game and are used to project future game states, but students do not interact directly with these models and simulations.

Impacts of Camps

Camps are often venues for games. Camps, defined broadly, have the following characteristics.

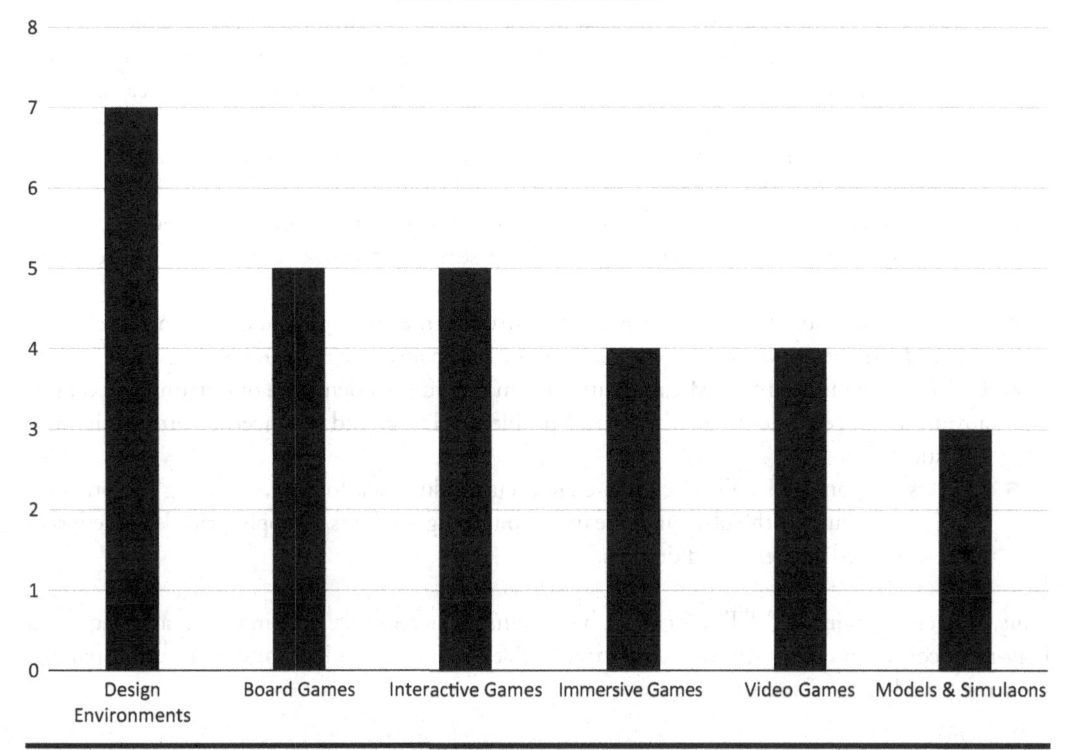

Figure 7.3 Games that enhance K–12 outcomes.

- STEM camps are offered by community groups, industry, public libraries, individuals reporting as Limited Liability Corporations (LLCs) and higher education institutions.
- STEM camps do not have a national or regional accrediting agency to evaluate curriculum, teacher expertise, or student learning outcomes for effectiveness and appropriateness.
- It is a challenge to determine the effectiveness of STEM camps as organizers fail to design quality assessment strategies. Most camp organizers settle for pre- and post-STEM camp attitudinal surveys.
- STEM camps are offered once a year as one-day, one week, or five-week sessions, increasing the difficulty of measuring student learning. Unlike traditional schooling that has grade progression, making it easier to assess student learning, the lack of STEM camp continuity contributes to the discussion of whether STEM camps can increase the K–12 STEM pipeline.

It is reasonable to assume students participating in STEM camps have different levels of literacy and numeracy skills and students participate for different reasons. Some participate at the urging of their parents to learn about STEM concepts and applications and some students purely for the socialization and fun aspect. Mixing students with different expectations and abilities at the same camp session reduces any chance to measure camp effectiveness.

A special class of camps is organized and operated by universities to prepare students to matriculate at their institutions. An analysis of university-sponsored STEM camps reveals common characteristics.

- University-sponsored STEM camps are utilized as a pre-college program to recruit future students to the university with the goal for the students to pursue a STEM major.
- University-sponsored STEM camps have a competitive admission and application process requiring a math or science teacher recommendation and school transcript indicating an A average in math and/or science courses.
- University-sponsored STEM camps are expensive, though the university provides some scholarship aid to assist students and families.
- University faculty serves as primary instructors supported by current students majoring in STEM fields. Faculty and college students serve in the role as mentors to the camp participants.
- University-sponsored STEM camps are designed for entering high school students, providing the students the opportunity to participate for four consecutive summers.
- University-sponsored STEM camp curriculum provides students the opportunity to master advanced concepts, logic, mathematical problem solving, and engineering probability and statistics.
- University-sponsored STEM camps enable curriculum scaffolding, building on previous years' curriculum with subsequent years encouraging students to apply prior knowledge to more advanced concepts and designs.

Thus, university-sponsored STEM camps have significant entrance requirements and require a long-term commitment by students and parents. Well-prepared students and committed parents are strong predictors of success.

Summary

There is a rich panorama of possible game-based interventions for increasing the number of K–12 students who are prepared for higher education. Much of the literature suffers from being mainly surveys of opinions. These findings are valuable but we have been looking for evidence of what works, not just whether students and teachers liked the experience.

We need data, information, and knowledge comparable to what we have for other domains, such as healthcare. We know the impacts for various health interventions on longevity. Our knowledge of the impacts of specific K–12 interventions on lifetime employment and income is scarce, to say the least. Of course, every health intervention is recorded to assure payments to providers. The situation in education is not at all similar.

Investments should focus on K–12 interventions not needing approvals of school boards, school administrators, or teacher unions. Well-designed games can be developed, deployed, and evaluated without such approvals. Games, camps, etc. should be carefully designed to achieve specific outcomes – and these outcomes should be assessed.

Here are six steps to successful design. Unfortunately, steps 5 and 6 are rarely conducted in education.

1. Define tasks of interest
2. Determine knowledge and skills needed to perform these tasks
3. Design interventions to foster these knowledge and skills
4. Design training environment to deliver interventions
5. Develop and evaluate the effectiveness of the environment
6. Assess effectiveness of training transfer to real-life jobs

College Students

Vlachopoulos and Makri (2017) provide a systematic review of the effect of games and simulations on higher education. They begin with a thorough review of earlier reviews. They then focus on 123 studies relevant to the effect of games and simulations on higher education. Findings relative to the learning outcomes of games and simulations are organized into three categories, namely cognitive, behavioral, and affective outcomes.

Cognitive outcomes refer "to the knowledge structures relevant to perceiving games as artifacts for linking knowledge-oriented activities with cognitive outcomes. "Serious gaming, especially given the context of enthusiastic students, has proved to be an effective training method in domains such as medical education, for example, in clinical decision-making and patient interaction." Other studies confirm the power of games and simulations in developing cognitive abilities, especially in instances of virtual simulations enhancing complex cognitive skills, such as self-assessment or higher-order thinking.

Behavioral objectives for higher education students refer to the enhancement of teamwork and improvement in relational abilities, as well as stronger organizational skills, adaptability, and the ability to resolve conflicts. "Simulation games are often seen as powerful tools in promoting teamwork and team dynamics, collaboration, social and emotional skills, and other soft skills, including project management, self-reflection, and leadership skills, which are acquired through reality-based scenarios with action-oriented activities."

Affective outcomes of using games and simulations in the learning process address student engagement, motivation, and satisfaction. In this area, "There seems to be a lack of shared definitions or taxonomy necessary for a common classification, which, therefore, results in terminological ambiguity."

Real-World Phenomena

A common way to introduce students to real-world phenomena is with simulations, which in some cases are physical, but more likely computational. In engineering, manufacturing simulations are common and portray the flow of materials to assembly of finished products.

In business, supply chain simulations are common, often portraying the flow of finished products from factories to distributors, wholesalers, and retailers. The MIT Beer Game, described below, is a good example, although this game is more often played by professionals rather than students.

SimCity is a great illustration that helps players to understand the dynamics of urban development, particularly how taxes and development projects influence public opinion. The public in *SimCity* wants a new stadium, for example, but resents tax increases to fund the project.

Environmental and weather simulations are both educational and useful for anticipating consequences of events such as hurricanes. One of my colleagues developed a storm surge model that accurately predicted water levels across Hoboken, NJ, as Hurricane Sandy caused the Hudson River to inundate the city in 2012.

Projections from weather simulations are often incorporated into weather forecasts. In some cases, multiple weather simulations are employed to compare forecasts based on differing sets of assumptions. The goal is not to pick the best answer, but to understand the range of possible futures.

Work Tasks and Practices

The *Health Advisor* game, as mentioned in earlier chapters, was developed and evaluated to train pre-med students at Emory University (Basole, Bodner, & Rouse, 2013). Students interviewed patient avatars and provided them guidance on appropriate next steps – referrals – given their symptoms. Players are advisors, not clinicians. Thus, the avatars are clients rather than patients in a traditional clinical sense.

At the start of game play, players pass through the lobby (Figure 7.4), enter their office (Figure 7.5), and choose whether to see a client, or to review client records, provider information, or their current performance score. During an appointment, the advisor interacts with the client, asks relevant questions, accesses the client's electronic health record (EHR), obtains relevant medical information through Medfile (an online medical information source), and based on assessment of the client's condition, makes a referral to a provider. Tests, treatments, and disease progression occur outside the health advisor's "game world" and are performed in the background by the simulation engine. Clients return to the health advisor after a certain period of time for a follow-up meeting.

The game has two classes of non-player characters. The first consists of the clients whom the advisors serve. The second consists of the providers to whom the advisors refer clients. The advisors have direct interaction with clients in that they see clients and have dialogue with them. The advisors do not have direct interaction with providers.

Our aim was to create a large set of clients with a broad range of characteristics, including age, gender, race, lifestyle, and disease. Clients are assumed to be from the general US adult population. Consequently, clients are sick in proportion to national morbidity rates. Each client has a specific disease and severity level. The assigned disease had to be representative of the age, race, and gender of the client. Clients were also assigned a set of lifestyle characteristics, which includes different levels of diet, exercise, and stress, which could impact the advisor's assessment and referral decisions. To ensure further game realism, each client was given a first and last name. Names were selected based on the age (i.e., traditional names were used for older clients), race, and gender of the client. We also used facial images for each client and ensured it matched the client demographic. Our final population contained 433 clients.

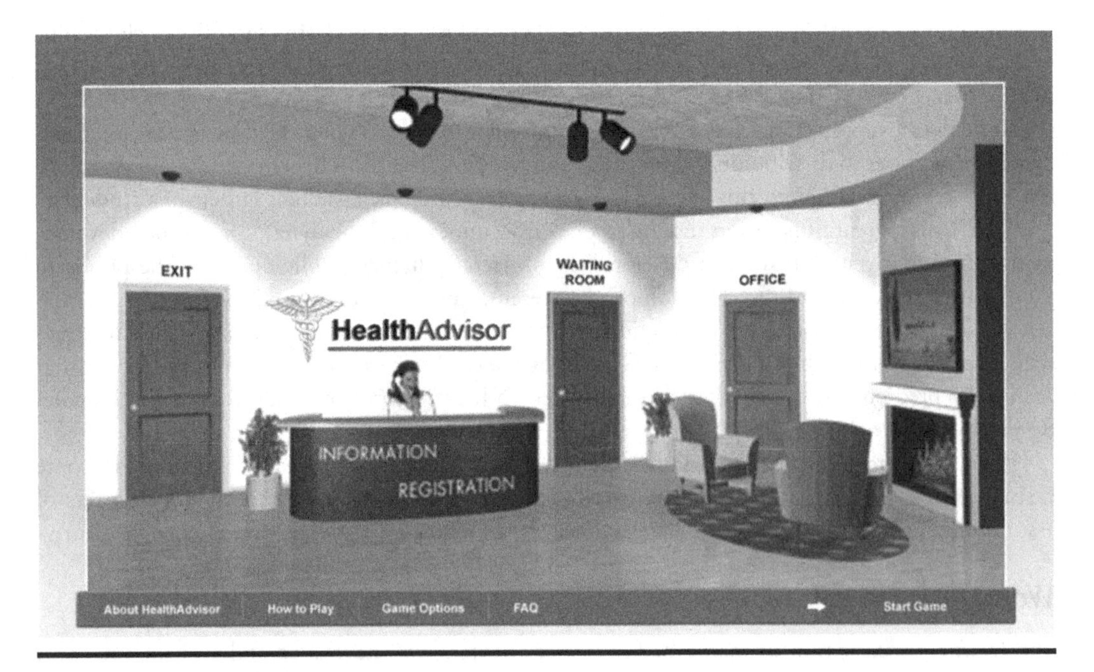

Figure 7.4 Lobby of *Health Advisor* game.

Figure 7.5 Office of *Health Advisor*.

A client can have one of any of 19 diseases. These diseases include breast cancer, colon cancer, lung cancer, prostate cancer, coronary artery disease, arrhythmias, heart valve disease, heart failure, stroke, bronchitis, emphysema, asthma, diabetes (type 1 and type 2), influenza, pneumonia, Alzheimer's, nephritis, and septicemia. Diseases have severity levels such as asymptomatic, symptomatic, chronic, and acute. Severity levels change in the course of time due to (a) natural disease progression or (b) the effect of treatment interventions. A severity level transition model was developed. The arcs represent the transition probabilities between severity states. Note that it is possible for two transitions to be possible from a given state (e.g., the transition from chronic to acute or recovered). The probabilities can be set to match a particular disease's progression pattern and transition probabilities. Thus, an acute disease may have positive transition probabilities only on those arcs that do not connect with the chronic severity state. A chronic disease, on the other hand, is likely to have positive transition probabilities on all arcs.

Disease progression is assumed to occur as a Markovian process, whereby for any given disease, a client may experience a transition from one severity to another at discrete points in time. A client who is asymptomatic with heart disease, for example, has a certain probability of transitioning from that severity to symptomatic. Transition probabilities are compiled from various medical studies sources, including feedback from experts. It should be noted that the fidelity and accuracy of transition probabilities are rough estimates. Exact transition probabilities are very difficult to estimate without considerable additional health information. Within the game, transitions occur at periodic game status updates (i.e., the transition points), and are executed by the simulation engine. The transition probability remains the same at each transition point. Thus, over time, the cumulative probability of transition increases. Transition paths are determined by Monte Carlo sampling.

Both health advisors and providers do not know a client's true health state, but rather must diagnose it based on symptoms. Thus, each client has a symptom state. The symptom state is modeled as a pair of disease (cause) and symptom. For each of the aforementioned diseases and their severity levels, we therefore have a set of symptoms. These symptoms are based on medical knowledge. For any disease not represented in a client's set of health state pairs, the client is asymptomatic. Some symptoms may be experienced in multiple diseases and a client may also have symptoms not related to their actual disease and severity level. Thus, a health advisor faces potentially significant challenges in assessing the disease of a client for treatment recommendations.

We considered a number of important issues in the design of the assessment dialogue between players and clients. Our first goal was to ensure that the content of clients' statements roughly reflects ground truth, although clients do not fully know their health states. Similarly, the content of clients' statements should reflect their previous treatments, the specific providers they have seen, and past conversations with the player. Furthermore, the health state of the client frames the symptoms that they will divulge – if probed by the player. The ways in which they state the experienced symptoms also depend on previous dialogue with the player – in both current and past meetings – and their personality. We thus created emotional state-based characters. As a result, clients' statements are modeled to be neutral, nervous, confident, know-it-all, passive, or collaborative.

The general setting of a client–advisor dialogue is across the player's desk (Figure 7.6). Dialogues begin with an introduction by the health advisor (e.g., "Hello" or "How are you doing?"). The dialogue then proceeds as a series of questions posed by the advisor and answers offered by the client. Questions posed by the advisor relate to symptoms, previous diagnoses, risk factors, payment, and treatments. Client response will depend on whether this is the first meeting or a follow-up meeting. In order to ensure consistency and dialogue logic, we developed a "client-advisor dialogue" state as well as a rule-set to determine allowable transitions between states. This capability allowed us to keep track of the course of an appointment and responses from previous ones. This is

Figure 7.6 Beginning of client interview.

important as it ensured, for example, that if the same question was asked twice during an appointment, a client's response would reflect that.

One of the central information sources of the game is the EHR. An EHR is a longitudinal electronic record of patient health information generated by one or more encounters in any care delivery setting, including patient demographics, progress notes, problems, medications, vital signs, past medical history, immunizations, laboratory data, and radiology reports. EHRs are major, complex software systems that capture massive amounts of clinical information. An inclusion of all possible fields is therefore beyond the scope of the game. Since there is no standard for EHRs, our design is tailored to support basic functionality for the game. The EHR captures mainly information from appointments with the health advisor and the results and outcomes of the tests and treatments conducted by providers. One of the unique aspects of our EHR is that it is automatically populated with relevant dialog information. In other words, once the player and client discuss something, it is captured automatically in the EHR.

At any point in the game, the advisor also has also access to an online medical knowledge base called Medfile. Medfile contains descriptive information on symptoms, diseases, diagnostic tests, treatments, and provider specialties. For each symptom, Medfile provides a description as well as the corresponding set of diseases and disease severity levels in which they are observed. For each disease, it lists a description including the most common symptoms, a list of confirming tests, and a list of possible treatments. Information in Medfile is organized by search category; within a category information is sorted alphabetically. We utilized a combination of reputable online medical resources to build Medfile (e.g., WebMD).

A key observation of game play is that Medfile access significantly increased both players' assessment and referral performance. This result suggests that information in Medfile helped players understand the symptoms, conditions, and referral options. Interestingly, Medfile was accessed most frequently for chronic cases. A possible explanation for this is that chronic diseases have a higher number of symptoms as well as overlapping symptoms with other diseases and therefore a greater understanding of the underlying disease roots is required. Over the course of play, Medfile access increased significantly, suggesting that players realized the value of having access to medical knowledge: 26% of players accessed Medfile for the first client, while 55% of players accessed Medfile for the last client.

Professional Practices

Teaching about professional practices is a primary objective of capstone projects, often referred to as senior design. I have advised many senior design teams, often involving projects related to healthcare delivery. These assignments often involved solving specific problems for the healthcare providers involved – the clients.

One project focused on emergency room waiting times at one of Emory Health's hospitals. The average waiting time had increased by one hour. Emory wanted to understand the source of this increase and what they could do about it. The student team used simulation to portray patient flow and determined that human factor deficiencies associated with a new software system were the culprit.

Another project involved bed turnover times at Piedmont Hospital. Patients were still in their rooms several hours after being supposedly discharged. Part of the solution was the creation of an outbound waiting room where discharged patients could wait for transportation home. Another cause of the delays was nurses batching the discharges and entering them all at the end of their shifts.

A third project involved scheduling operating rooms (ORs) at Piedmont Hospital. The prevailing process was for each surgeon to designate their desired time and preferred OR. In the process of trying to formulate the optimal scheduling algorithm, the student team identified two useful heuristics.

First, surgeons should only be able to designate their preferred half day. Second, if the surgeon is one who frequently runs over their scheduled time, they have to be at the end of the half day period. The students applied these heuristics to a recent year's surgery records and determined performance could be improved with one less OR, which enabled management to avoid investing in adding an OR.

All three projects had similar impacts on the student teams. At first, they all were overwhelmed by the complexities of their assignments. Beyond the technical issues at hand, all three problems were laced with behavioral and social phenomena. They worked to characterize these phenomena and understand their impacts. Fortunately, they all succeeded and were quite proud of their findings and recommendations.

Employees

Games can greatly enhance employee training. Domain-specific work practices can be demonstrated and learned. Real-world phenomena can be experienced and better understood. Elements of business strategies and approaches to addressing them can be understood and practiced.

Domain-Specific Work Practices

Many work practices are learned by osmosis. This unconscious assimilation of ideas, knowledge, and technical skills can be effective, but is often very inefficient. This is particularly true when scarce resources are involved and/or when safety considerations are central.

In Chapter 4, I mentioned the use of lower-fidelity training simulators to prepare trainees for immersion in much higher-fidelity, full-scope training simulators. Marine Safety International's high-fidelity supertanker engine room simulator only allowed for one trainee at a time. This reflects the industry's practice of having a single engineer on duty during each shift.

Beyond this capacity constraint, the objective of the lower-fidelity simulators was to prepare trainees to take advantage of their limited time in the high-fidelity simulator. Thus, the lower-fidelity simulators introduced them to terminology and relationships that were highly relevant to the scenarios they would encounter and address in the high-fidelity simulator.

We studied the training program for the Navy's Aegis Combat Information Center, which is located in Moorestown, NJ, and visible from the New Jersey Turnpike. The building was designed to resemble an actual Aegis cruiser, and is referred to as the Cornfield Cruiser.

The training includes half-day exercises involving the full complement of 25 Navy crew members in a high-fidelity full-scope simulator. We focused on the five-person anti-air team. We learned that these personnel had rather deficient team mental models as discussed in Chapter 4.

This led us to develop a lower-fidelity Team Model Trainer (TMT) to support trainees prior to their engagement with the high-fidelity simulator. The TMT addressed the information needed by each of the five team members, as well as the information they could be counted on to provide. Each team member learned the roles of the other members of their team. This greatly enhanced their performance in the high-fidelity simulator.

Real-World Phenomena

Didactic teacher-centered instruction in which teachers deliver and students receive lessons is often a good starting point. However, such introductory material is often best followed by being immersed in real-world phenomena. This leads to experiential learning. Games can be an invaluable means for such learning.

The MIT Beer Game (Sterman, 1992) introduces players to the structure and dynamics of supply chains. In the following description of the game, I have underlined factors that make the game surprisingly difficult.

> The game is played on a board that portrays the production and distribution of beer. Each team consists of four sectors: Retailer, Wholesaler, Distributor, and Factory arranged in a linear distribution chain. One or two people manage each sector. Pennies stand for cases of beer. A deck of cards represents customer demand. Each simulated week, customers purchase from the retailer, who ships the beer requested out of inventory. The retailer in turn orders from the wholesaler, who ships the beer requested out of their own inventory. Likewise the wholesaler orders and receives beer from the distributor, who in turn orders and receives beer from the factory, where the beer is brewed. At each stage there are shipping delays and order processing delays. The players' objective is to minimize total team costs. Inventory holding costs are $.50/case/week. Backlog costs are $1.00/case/week, to capture both the lost revenue and the ill will a stockout causes among customers. Costs are assessed at each link of the distribution chain.
>
> The game can be played with anywhere from four to hundreds of people. Each person is asked to bet $1, with the pot going to the team with the lowest total costs, winner take all. The game is initialized in equilibrium. Each inventory contains 12 cases and initial throughput is four cases per week. In the first few weeks of the game the players learn the mechanics of filling orders, recording inventory, etc. During this time customer demand remains constant at four cases per week, and each player is directed to order four cases, maintaining the equilibrium. (Customer demand begins at four cases per week, then rises to eight cases per week in week five and remains completely constant ever after.) Beginning with week four the players are allowed to order any quantity they wish, and are told that customer demand may vary; one of their jobs is to forecast demand.
>
> Each player has good local information but severely limited global information. Players keep records of their inventory, backlog and orders placed with their supplier each week. However, people are directed not to communicate with one another; information is passed through orders and shipments. Customer demand is not known to any of the players in advance. Only the retailers discover customer demand as the game proceeds. The others learn only what their own customer orders. These information limitations imply that the players are unable to coordinate their decisions or jointly plan strategy, even though the objective of each team is to minimize total costs. As in many real life settings, the global optimization problem must be factored into sub-problems distributed throughout the organization.
>
> When customer orders increase unexpectedly, retail inventories fall, since the shipment delays mean deliveries continue for several weeks at the old, lower rate. Faced

> with a growing backlog, people must order more than demand, often trying to fix the problem quickly by placing huge orders. These large orders stock out the wholesaler. Retailers don't receive the beer they ordered, and grow increasingly anxious as their backlog worsens, leading them to order still more, even though the supply pipeline contains more than enough. Thus the small step in demand from four to eight is amplified and distorted as it is passed to the wholesaler, who reacting in kind, further amplifies the signal as it goes up the chain to the factory.
>
> Most people do not account well for the impact of their own decisions on their teammates – on the system as a whole. In particular, people have great difficulty appreciating the multiple feedback loops, time delays and nonlinearities in the system, using instead a very simple heuristic to place orders.

Delays, uncertainties, and information limitations aggravate these very human inclinations. Of course, once players come to understand these real-world phenomena, they can better appreciate the benefits of modern supply chain management systems that provide global transparency of the state of the system across stages.

Elements of Competitive Strategy

As indicated in Table 7.1, senior employees also need to learn. They may have become Vice President due to their superior technical skills in finance, engineering, or marketing, but their strategic thinking skills often need a significant upgrade. This upgrade can be abetted via game-based exploration of alternative futures and formulation of plans for addressing these futures.

We developed a series of four software tools that comprised the *Advisor Series* of planning tools, summarized in Figure 7.7 (Rouse, 2007, 2019). These tools included:

- *Situation Assessment Advisor* (SAA)
- *Business Planning Advisor* (BPA)

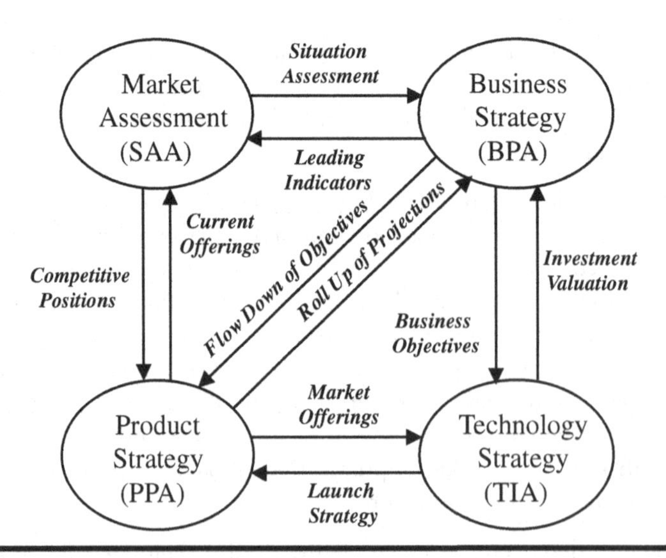

Figure 7.7 The *Advisor Series* of planning tools.

- *Product Planning Advisor* (PPA)
- *Technology Investment Advisor* (TIA)

The basic idea was that companies would use SAA to assess their market situation, which would inform using BPA to formulate their business strategy, which would drive their use of PPA to plan product and service offerings, which would be informed by the technology investment portfolio formulated with TIA to enable these offerings.

This provided a pretty compelling marketing story, but PPA and TIA dominated our engagements with clients. This was due, at least in part, to these tools being more sophisticated in terms of explicit models rather than more opaque expert systems. In other words, users created models with PPA and TIA that they could then use to explore various scenarios. With SAA and BPA, the underlying models were "hard wired."

We envisioned the flows between the four tools depicted in Figure 7.7 as being automatic, which would be difficult to accomplish seamlessly. Fortunately, we never attempted this because customers had a better idea. They were all quite facile with Microsoft Excel. They would capture model output in Excel and then paste it into another model, with perhaps a bit of translation.

The reason they preferred this is that the Excel spreadsheet then became the "minutes" of the working session. Their confidence and expertise with Excel seemed to provide our tools more credibility. The ability to use their preferred representation, i.e., a spreadsheet, increased their level of comfort with the whole process depicted in Figure 7.7. This is an important aspect of acceptability.

Customers provided another important insight. The last step in the menu for each tool is Generate Documentation. Use of this step would automatically create a set of slides for use with Microsoft PowerPoint. This slide set would capture whatever elements of the product plan the users chose, via a dashboard designed for this. While customers asked for this function within our tools, they seldom used it.

A marketing executive at one of our clients explained this lack of use. He said that he used this function and brought the slide deck to a meeting with his boss for the purpose of requesting a budget to proceed. His boss asked how he had created the slides, and the marketing executive opened his laptop and showed him PPA. His boss became intrigued, played around with various assumptions, considered alternative scenarios, and approved his budget. From then on, he always presented the tool itself, never the slides.

We have experienced this phenomenon repeatedly. When participants in PPA or TIA sessions can take the controls and explore possible futures, their buy-in soars. They also become a rich source of ideas for improving models and visualizations. I think they were also tired of looking at PowerPoint slides.

How does the *Advisor Series* qualify as a game-based approach to learning? Quite simply, the predominant use of these tools was in "What if?" exercises. They were exploring alternate futures for aircraft engines, automobile sensors, batteries, optical computing, satellites, semiconductors, software, wireless networks, etc. Users included Boeing, Digital, Honeywell, Hughes, Johnson Controls, Lockheed, 3M, Motorola, NCR, Raytheon, Rival, Rolls Royce, Rover, and government sponsors.

To introduce users to this approach to exploration, we devised a training scenario called Grocery Buyers Assistant. The goal of the players was to research, develop, and deploy a computer-based decision-support system for consumers in grocery stores. We chose this context for two reasons. First, all of our users understood the idea of buying groceries.

Second, none of our users was even remotely involved in the retail grocery industry. Thus, they did not second guess any of the background information that we provided. They simply played the game of pretending that they were in that business. They focused on the use of the planning tools to formulate market offerings, address trade-offs, and foster acceptable economic returns.

The overall approach supported by the *Advisor Series* led to many major investments by clients, numerous market successes, and a few home runs. Of course, we were learning as well. Many of our lessons learned about these companies and their markets are summarized elsewhere (Rouse, 1996, 1998, 2007).

Key Enablers

Pedagogy

The method and practice of teaching need to morph away from teacher-centric didactic presentations. Experiential learning should be central. Teachers need to move away from being "Sages on Stages" to become "Guides on the Sides," helping students to progress in game playing and leverage what they are learning to achieve higher levels of performance. This may be a significant challenge for many traditional teachers.

Curriculum

The curriculum is defined by the subjects comprising a course of study. This will, of course, be highly influenced by the domain of study. However, all curricula should include material on game playing, elements of collaboration, and leadership. There should also be an emphasis on the measurement of learning outcomes, including assessments of learning transfer beyond educational settings.

Technology

Not all games are technology enabled, but computing technology is increasingly pervasive. Here are several areas where technology trends are particularly strong and need to be exploited.

Online Education. All disciplines will necessarily have to entertain greater use of online teaching as the response to the pandemic has prompted. However, disciplines may differ in emphases. Some of these differences will be driven by the differing content employed in the curricula of these disciplines. Also of great importance, will be the extent that face-to-face interactions are central to each discipline and the extent to which these interactions can be technologically mediated.

Interactive Technologies. Advanced technology can enable compelling interactive portrayals of phenomena ranging from chemistry and physics, to human physiology and behaviors, to social and cultural interactions. These interactive technologies can augment reality and provide profound educational experiences. The quality of these immersive portrayals has steadily improved and the costs, at least on widely available platforms, have progressively decreased. The economics of such technologies depend, however, on the number of students across which costs can be amortized.

Knowledge Management. Information access and knowledge management are challenges across disciplines, although the nature of data and knowledge artifacts differ substantially across disciplines. In particular, the technological infrastructure associated with science and technology has benefited from enormous investments. Humanities have seen important investments and innovations but not at all on the same scale. Of particular note, the data and knowledge artifacts of the humanities were seldom originally created digitally.

Process Improvement. Process modeling and improvement initiatives are significantly affected by two factors. One is the extent to which educational processes are interwoven with operational processes. This is greatest for medicine where much of education happens during delivery of clinical services. In engineering, considerable research happens with industry and undergraduate cooperative education programs are pervasive. Humanities have few similar processes and thus can be approached in a more straightforward manner.

The second factor is scale. When an undergraduate major, e.g., electrical, industrial, or mechanical engineering, has well over 1,000 students in one department, technology investments can be amortized across many students and, thereby, justify much greater investments. If such institutions are also well resourced, the human and financial resources can be marshaled to undertake these investments.

Conclusions

This chapter has focused on game-based approaches to pursuing learning objectives for K–12 students, college students, and employees. I summarized a wealth of research findings and provided detailed illustrations of numerous games.

In the next chapter I address serious games. Clark Abt is credited with coining the term "serious games" in the 1970s, defined as games that have an "explicit and carefully thought-out educational purpose and are not intended to be played primarily for amusement." Abt also recognized that this "does not mean that serious games are not, or should not be, entertaining" (Djaouti et al., 2011).

In particular, Chapter 8 addresses game-based approaches to pursuing real-world complex problems in healthcare, education, and transportation. These problems and their resolution are elaborated in considerable detail.

References

Agbo, F.J., Oyelere, S.S., Suhonen, J., & Tukiainen, M. (2022). Design, development, and evaluation of a virtual reality game-based application to support computational thinking. *Educational Technology Research and Development*, https://doi.org/10.1007/s11423-022-10161-5.

Al-Zaytoonah, M.H.A-T (2016). The effectiveness of educational games on scientific concepts acquisition in first grade students in science. *Journal of Education and Practice*, 7 (3), 31–37.

Alemân, M.W., Tomko, M.E., Linsey, J.S., & Nagel, R.L. (2022). How do you play that makerspace game? An ethnographic exploration of the habitus of engineering makerspaces. *Research in Engineering Design*, 33, 351–366, https://doi.org/10.1007/s00163-022-00393-0.

Arztmann, M., Hornstra, L., Jeuring, J., & Kester, L. (2022). Effects of games in STEM education: A meta-analysis on the moderating role of student background characteristics, *Studies in Science Education*, https://doi.org/10.1080/03057267.2022.2057732.

Basole, R.C., Bodner, D.A., & Rouse, W.B. (2013). Healthcare management through organizational simulation. *Decision Support Systems*, 55, 552–563.

Bertram, L. (2020). Digital learning games for mathematics and computer science education: The need for preregistered RCTs, standardized methodology, and advanced technology. *Frontiers in Psychology*, https://doi.org/10.3389/fpsyg.2020.0212.

Djaouti, D., Alvarez, J., Jessel, J-P., & Rampnoux, O. (2011). Origins of serious games. *Serious Games and Edutainment Applications*. 25–43. https://doi.org/10.1007/978-1-4471-2161-9_3. ISBN 978-1-4471-2160-2.

Elme, L., Jørgensen, M.L.M., Dandanell, G., Mottelson, A., & Makransky, G. (2022). Immersive virtual reality in STEM: Is IVR an effective learning medium and does adding self-explanation after a lesson improve learning outcomes? *Educational Technology Research & Development*, 70, 1601–1626. https://doi.org/10.1007/s11423-022-10139-3.

Jong, M,S-Y. (2022). Flipped classroom: motivational affordances of spherical video-based immersive virtual reality in support of pre- lecture individual learning in pre-service teacher education. *Journal of Computing in Higher Education*, https://doi.org/10.1007/s12528-022-09334-1.

LiI, E., Lam, A.T., Fuhrmann,T., Erikson, L., Wirth, M., Miller, M.L., Blikstein, P., & Riedel-Kruse, I.H. (2022). DIY liquid handling robots for integrated STEM education and life science research, *PLoS One*, 17 (11), e0275688.

Rouse, W.B. (1996) *Start Where You Are: Matching Your Strategy to Your Marketplace*. San Francisco, CA: Jossey-Bass.

Rouse, W.B. (1998). *Don't Jump to Solutions: Thirteen Delusions That Undermine Strategic Thinking*. San Francisco, CA: Jossey-Bass.

Rouse, W.B. (2007). *People and Organizations: Explorations of Human-Centered Design*. New York: Wiley.

Rouse, W.B. (2019). *Computing Possible Futures: Model Based Explorations of "What If?"* Oxford, UK: Oxford University Press.

Rouse, W.B., Lombardi, J.V., & Gargano, M. (2023). *Policy Innovations To Enhance The STEM Talent Pipeline: Interventions to Increase STEM Readiness of K-12 Students*. Hoboken, NJ: Systems Engineering Research Center, Stevens Institute of Technology.

Sterman, J. D. (1992). Teaching takes off: Flight simulators for management education. *OR/MS Today*, 19 (5), 40–44.

Vlachopoulos, D., & Makri, A. (2017). The effect of games and simulations on higher education: A systematic literature review. *Technology in Higher Education*, 14 (22). https://doi.org/10.1186/s41239-017-0062-1.

Wang, L-H. Chen, Hwang, G-J., Guan, J.Q., & Wang, Y-Q. (2022). Effects of digital game-based STEM education on students' learning achievement: a meta-analysis, *International Journal of STEM Education*, 9, 26. https://doi.org/10.1186/s40594-022-00344-0.

Zhang, S., Wong, G.K.W., & Chan, P.C.F. (2022). Playing coding games to learn computational thinking: What motivates students to use this tool at home? *Education and Information Technologies*, https://doi.org/10.1007/s10639-022-11181-7.

Chapter 8

Serious Games

Introduction

Serious games are designed for primary purposes other than pure entertainment. The idea shares aspects with simulation in general, but explicitly adds elements of fun and competition. This chapter reviews three serious games in detail in terms of development, deployment, and impacts. These games include applications addressing impacts of the Affordable Care Act on the New York City health ecosystem, economic impacts of alternative futures on research universities, and impacts of Battery Electric Vehicles (BEVs) and Autonomous Vehicles (AVs) on the automobile and insurance industries.

Computational models can provide the basis for an immersive interactive game environment as shown in Figure 8.1. This *Immersion Lab* includes seven touch-sensitive displays arranged in an 8 by 20 foot 180 degree semicircle. Thus, users can enter into the complexity of their world.

Off-the-shelf software enables one to configure the displays into arrangements ranging from one to seven independent displays. Each display can host one or more models, typically statistical models or computational simulations. In some cases, the models are interactive games. Typically, models on one screen are linked to models on other screens, providing inputs and/or outputs to each other. In this way, users can see how one portion of their world affects the other parts.

The use and value of the *Immersion Lab* are best elaborated with examples. In this chapter, I discuss illustrative games in healthcare, education, and energy. The references provided for each game discuss the computational models employed in substantial detail. My emphasis here is on the serious problems these games were developed to address. The game-oriented nature of these applications was designed to engage key stakeholders in exploration, discussion, and debate.

Healthcare

New York City Health Ecosystem

The Affordable Care Act (ACA) of 2010 has caused a transformation of the healthcare industry. As shown in Figure 8.2, this industry involves complicated relationships among patients, physicians, hospitals, health plans, pharmaceutical companies, healthcare equipment companies, and

DOI: 10.4324/9781003491927-8

Figure 8.1 ***Immersion Lab*** **for serious games.**

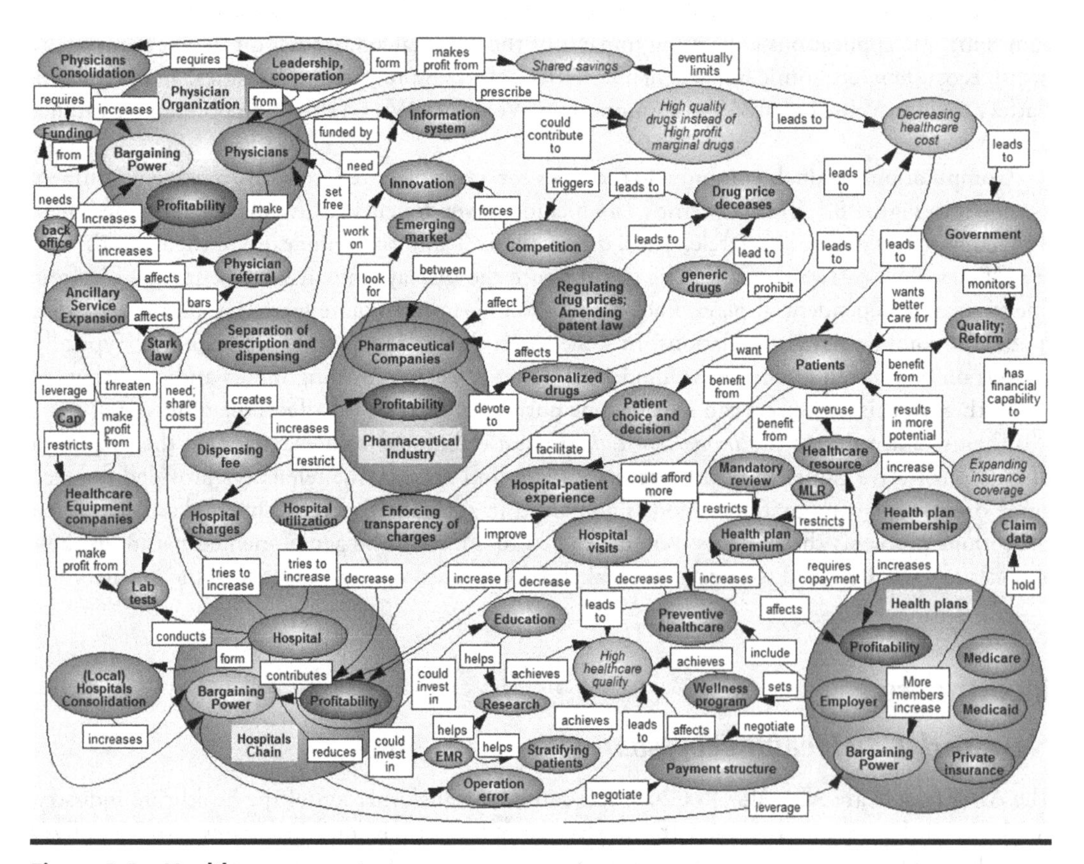

Figure 8.2 **Healthcare ecosystem.**

government. Hospitals are uncertain about how they should best respond to threats and opportunities. This is particularly relevant for hospitals located in competitive metropolitan areas such as New York City, where more than 50 hospitals are competing – many among the nation's best.

We conducted an interview study of healthcare provider executives. They reported the following concerns that have arisen in this uncertain environment:

- What if we wait until the healthcare market stabilizes and only invest in operational efficiency?
- Should we merge with competing hospitals to increase negotiation power?
- Shall we only focus on acquiring physician practices in highly reimbursed diagnostic groups?

We developed a data-rich agent-based simulation model to study dynamic interactions among healthcare systems in the context of merger and acquisition (M&A) decision making (Yu, Rouse, Serban, & Veral, 2016). By "rich" we meant extensive rule sets and information sources, compared to traditional agent-based models. The computational model included agents' revenues and profitability (i.e., financial statements), operational performance and resource utilization, as well as a more detailed set of objectives and decision-making rules to address a variety of what-if scenarios. The use of this model is summarized in Figure 8.3.

We applied our modeling approach to M&A dynamics of hospitals in New York City, informed by in-depth data on 66 hospital corporations of the Hospital Referral Region in the Bronx, Manhattan, and Eastern Long Island. The objective of the simulation model was to assist hospital executives to assess the impact of implementing strategic acquisition decisions at the system level. This was accomplished by simulating strategies and interactions based on real historical hospital balance sheets and operational performance data.

The outcomes of the simulation included the number of hospitals remaining in the market and frequent M&A pairs of hospitals under various settings. By varying strategy inputs and relevant parameters, the simulation was used to generate insights as to how these outcomes would change under different scenarios. The interactive visualizations shown in Figure 8.1 complemented the simulation model by allowing non-technical users to interactively explore relevant information, to input parameter values for different scenarios, as well as to view and validate the results of the simulation model.

The results from the simulation model facilitated M&A decision making, particularly in identifying desirable acquisition targets, aggressive and capable acquirers, and frequent acquirer–target pairs. The frequencies of prevalent pairs of acquirer and target appearing under different strategies in our simulation were of particular interest. The frequency level is a relative value in that it depends on the number of strategies included and hospitals involved. A high frequency suggests a better fit and also repeated attraction.

Validation of agent-based simulations is challenging, especially for high-level strategic decision simulations. The overall model and set of visualizations were validated in two ways. From a technical perspective, we compared our simulation results with Capital IQ's hospital mergers and acquisitions transaction data set. Although there was a limited number of cases under our regional constraint in Capital IQ's database, the realized M&A transactions appeared in our results.

Second is the feedback from users. There were many, roughly 30, demonstrations to hospital decision makers and healthcare consultants as well as senior executives from insurance, government, foundations, etc. In total, perhaps 200 people participated in the demos and many took the controls and tried various options. They made many suggestions and the number of types of interactive visualizations iteratively increased.

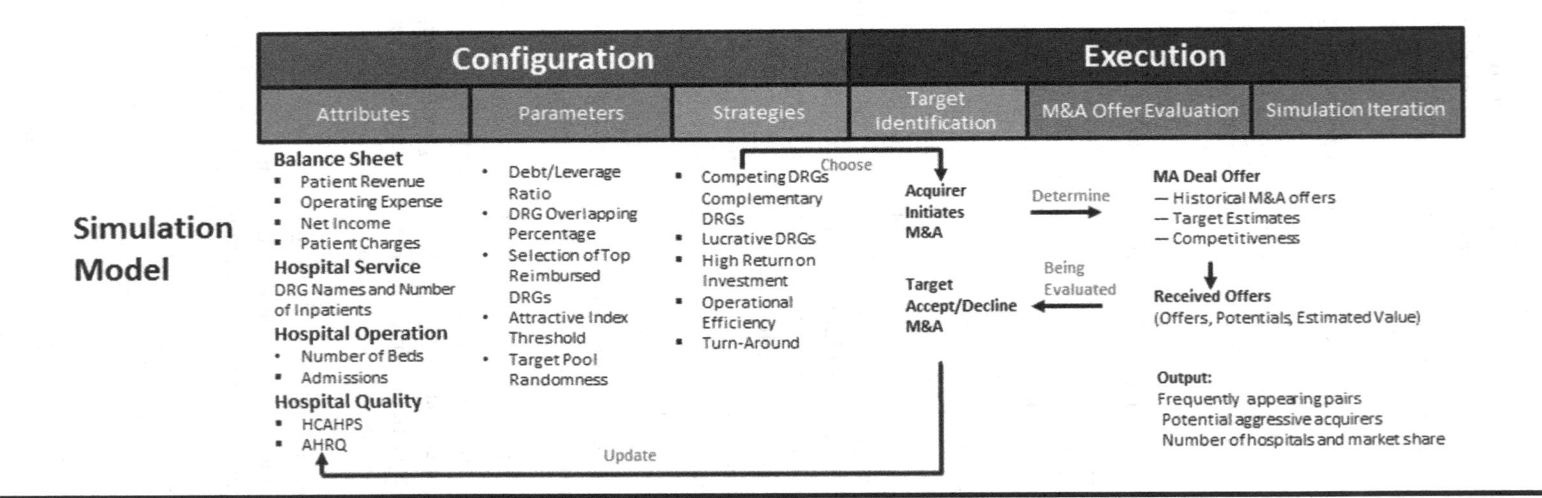

Figure 8.3 Configuring and executing New York City M&A model.

Two predictions were of particular interest. We correctly predicted the Mt Sinai acquisition of Beth Israel. We incorrectly predicted that Mt Sinai would acquire Staten Island University Hospital. Instead, Northwell acquired Staten Island. During a demo to Mt Sinai, we noted this prediction. A Mt Sinai executive said that Staten Island was at the top of their list, but Northwell acted more quickly. So, this prediction was not that far off.

The key value of the overall model and set of visualizations was, of course, the insights gained by the human users of this environment. For example, they could determine the conditions under which certain outcomes are likely. They can then monitor developments to see if such conditions are emerging. Thus, they know what *might* happen, even though they cannot be assured of what will happen. The greatest insights are gained not only from simulation but also from interactive visualizations that enabled massive data exploration of hospital performance in highly competitive diagnostic groups.

Washington DC Opioid Ecosystem

In the 1980s, pain increasingly became recognized as a problem that required adequate treatment. Pharma's business model for pain pills included misleading advertising, an aggressive sales force, and incentives for doctors to prescribe opioids. Once addicted, people found heroin to be much cheaper than prescription drugs, and more powerful.

Unfortunately, the service supply chain for treating substance abuse is highly fragmented. Before the present epidemic, opioids were prescribed mainly for short-term uses such as pain relief after surgery or for people with advanced cancer or other terminal conditions. But in the US, the idea that opioids might be safer and less addictive than was previously thought began to gain credibility.

Prescriptions for opioids increased gradually throughout the 1980s and early 1990s. In the mid-1990s, pharmaceutical companies introduced new opioid-based products. Of particular note is Purdue Pharma's OxyContin, a sustained-release formulation of oxycodone. Prescriptions surged and the use of opioids to treat chronic pain became widespread.

The structure of the healthcare system in the United States also contributed to the over-prescription of opioids. Because many doctors are in private practice, they can benefit financially by increasing the volume of patients that they see, as well as by ensuring patient satisfaction, which can incentivize the over-prescription of pain medication. Prescription opioids were also cheap in the short term. Patients' health insurance plans often covered pain medication but not pain-management approaches such as physical therapy.

I conducted a series of interviews with front-line health professionals at MedStar Health in Baltimore. Interviewees included nurses, social workers, and recovery coaches. My interview notes include litanies of needs for better integration of care across health, education, and social services.

One nurse complained of the enormous percentage of her time she spends on the phone trying to coordinate the services needed by patients. A social worker commented that substance abuse was typically just one of a patient's problems. Other problems included mental health challenges, joblessness, and homelessness. A recovery coach noted one patient who had been in the Emergency Department several hundred times over a two-year period.

There have been concerted efforts to limit opioid prescription in terms of both numbers of pills and refills. This will likely limit the growth of the number of new addicts. However, as noted earlier, it forces existing addicts to seek new and less safe sources of opioids and heroin. A much more integrated approach to population health is needed to assist these people.

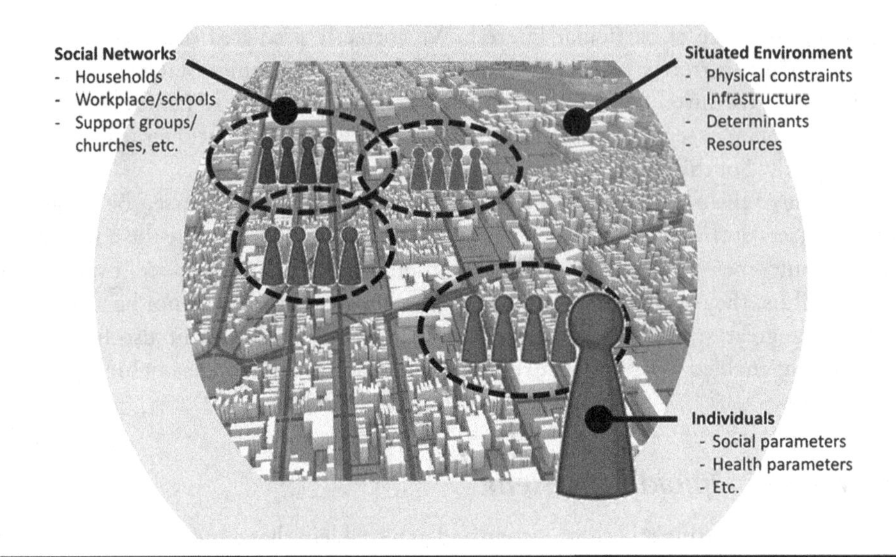

Figure 8.4 Artificial society simulation of Washington, DC.

Working with a research team at MITRE, we developed an agent-based model of the population of Washington, DC, at the level of each individual citizen, i.e., 700,000 agents (Tolk et al., 2023). Based on thorough research of the medical literature, we developed a state-based model of each agent that included comorbidities and social determinants of health. A range of data sets was employed to tailor the agent models to the residents of each of the city's eight wards (Figure 8.4).

Our primary interest was in the extent to which social interventions, rather than medical interventions, could motivate addicts to enter recovery. In particular, we were intrigued by the impacts of peer recovery coaches reported in the medical literature. We hypothesized that an addict's likelihood of seeking recovery is related to the number of recovered addicts in their social network.

We used the model to predict the number of emergency room visits due to overdoses and the number of deaths due to overdoses for each ward. Our predictions were quite close to the actual data, including the wide variations between the poorer and more affluent wards in Washington, DC. Unexpectedly – and unfortunately – we were able to use the model to conduct a natural experiment.

Our experiment involved predicting the impacts of social isolation, due to the coronavirus, on opioid abuse overdoses and deaths. We socially isolated the agents in our model. Thus, current addicts had no interactions with recovered addicts. The model predicted increases in overdoses and deaths. Our predictions were in the ballpark with what actually happened.

We also evaluated the impacts of a possible new vaccine that prevents opioid overdoses. This seems like a great idea but may backfire. If this vaccine greatly reduces the likelihood of overdoses, addicts may not be motivated to enter recovery. Thus, there will be fewer recovered addicts to convince current addicts to attempt recovery. The result may be more addicts.

Summary

These two model-based games enabled stakeholders to explore important policy options. Both were highly evidence-based, drawing upon extensive real-world data sets. The players of these games were fascinated by the availability of such data. For the most part, they could have accessed these data sets on their own, but did not. Of course, the models integrated multiple data sets in ways that required some effort. The models also employed this data within computational frameworks with which many users were unfamiliar until they played our games.

Higher Education

The Tennenbaum Institute (TI) at Georgia Tech, during my tenure as Executive Director, was focused on the aerospace, automotive, and healthcare industries. Members of the TI Advisory Board often asked me when I was going to address the transformation of higher education. I could not imagine anyone sponsoring that, so I soft-pedaled the topic. Then my Advisory Board at Stevens asked me the same question. I started paying attention.

I delved into higher education, trying to understand why education had replaced healthcare delivery as the poster child for runaway costs. Increases in the price of tuition and fees were far outstripping increasing healthcare costs and, by far, any increases in costs of living. My scrutiny of these phenomena resulted in *Universities as Complex Enterprises* (Rouse, 2016) and subsequent articles.

This topic could easily dominate the agendas of endless faculty meetings and retreats, with lots of strong opinions and limited progress. Faculty members from chemistry, history, and mathematics, for example, would articulate precisely what was wrong and what to do about it. Few, if any, faculty members would have any data beyond personal experiences.

My objective was to create an evidence-based computational environment where alternative futures could be explored, discussed, and debated, as typified by the environment depicted in Figure 8.1. This model-based game would enable disciplinary experts, now Provosts, Deans, and Department Chairs, to deeply understand the complex organizational system they were currently leading.

Four Scenarios

Research universities face enormous strategic challenges as the costs of education steadily increase far beyond increases in the cost of living. They also face increasing global competition. This section elaborates four likely scenarios for the future of research universities.

Scenario development should be based on best practices on this topic (Fahey & Randall, 1998; Schoemaker, 1995; Schwartz, 1991). All of the pundits begin by defining the forces that drive the future. There are – at least – four strong driving forces that will affect academia's future:

- Competition among top universities will become increasingly intense, both for talent and resources – there will be a clash of the titans
- Globalization will result in many academic institutions, particularly in Asia, achieving parity in the competition – it will become hot, flat, and crowded
- Demographic trends portend an aging, but active populace leading to an older student population – higher education will need to become a lifespan Mecca
- The generation of digital natives will come of age, go to college and enter the workforce – there will be no choice but to become a networked university

We cannot escape these forces; nor can we fully predict the ways in which they will interact to shape the world over the next couple of decades. We can be sure, however, that for academic institutions to compete in this future, their strategies must be sufficiently robust to accommodate these forces. If, instead, they focus on just one scenario – for example, the clash of titans that most closely resembles business as usual, perhaps on steroids – they will almost certainly be at a competitive disadvantage in the future.

Clash of Titans. I have worked at, consulted with, or served on advisory boards of quite a few top universities. Every one of them pays attention to their *US News & World Report* ranking. They aspire to battle with the titans of higher education, and hold their own. This scenario has universities continuing that clash, perhaps clawing their way to higher rankings, albeit in an increasingly competitive environment.

Hot, Flat, & Crowded. Tom Friedman (2005) has argued that the world is flat and we should no longer assume business as usual – his revision of this bestseller included a chapter on Georgia Tech and how they transformed education in computing. More recently, Friedman (2008) has argued that the world will be hot, flat, and crowded. In this scenario, academic institutions have to compete with a much wider range of players in a global arena.

Lifespan Mecca. It is easy – and convenient – to assume that the students of the future will be much like the students of today. However, CSGNET (2007) reports that over the past decade the number of graduate students 40 years old and older has reached record numbers. From 1995 to 2005, the number of post-baccalaureate students age 40 and older at US colleges and universities jumped 27%. And during the next two decades, the number of older citizens will rise at even faster rates than the number of those 24 and younger, which suggests that the number of post-baccalaureate students age 40 and over very likely will continue to grow. In this scenario, universities have to address a "student" population with more diverse interests and expectations rather different from students of the past and current eras.

Network U. Technology is increasingly enabling access to world-class content in terms of publications, lectures, and performances. Higher education can leverage this content to both increase quality and lower costs (Kamenetz, 2009; North, 2009). This technology has also spawned the generation of "digital natives" that is always connected, weaned on collaboration, and adept at multi-tasking. In this scenario, academia has to address different types of students using very different approaches to delivering education and conducting research.

Framing the future 20 years from now is quite difficult. Yet, this is essential if academic institutions are to focus their competencies and resources on the possible futures in which our students – and all of us – will have to compete. Our abilities to understand and manage the inherent uncertainties associated with these futures can be an enormous competitive advantage. We need to enhance these abilities to maintain our competitive position in global education.

Economic Model

We developed a computational model for predicting how various strategic decisions will affect research universities as they address the four scenarios. The goal was to enable university leaders to explore alternatives and trade-offs in their strategies for the future (Rouse, 2016). Many of these explorations were conducted in the *Immersion Lab*, with Provosts, Deans, and Department Chairs addressing a range of "What if?" questions, often unearthing underlying relationships of which they had been unaware.

Figure 8.5 portrays the overall flow of variables within the economic model. Not every connection is portrayed, as the figure would become hopelessly messy. Of particular note, students' applications and enrollments are driven by a trade-off between net tuition and brand value. Somewhat

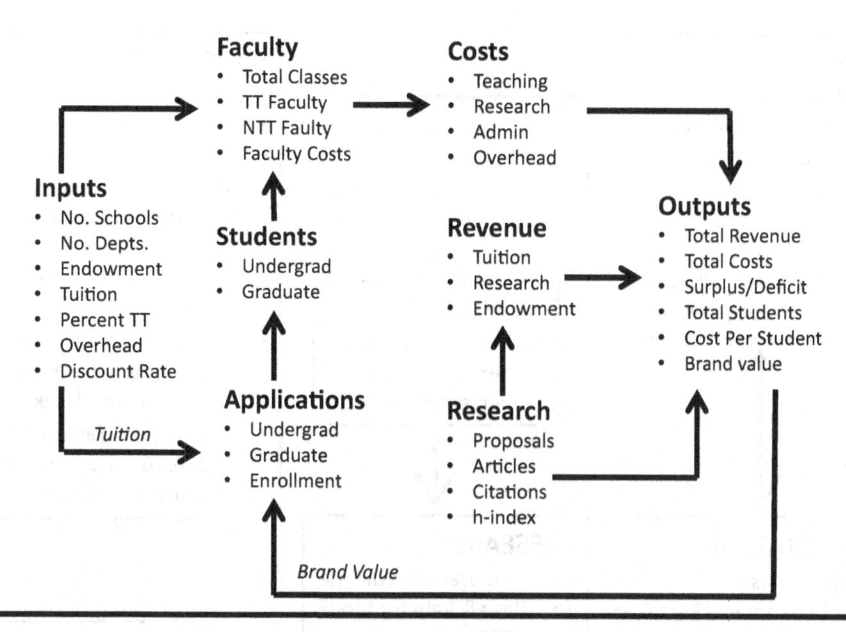

Figure 8.5 Overall structure of economic model of research universities.

simplistically, students seek to matriculate at the highest brand value university that they can afford.

Figure 8.6 shows how all the models come together, with the variables within each model listed. The financial model that follows Figure 8.5 is not shown in Figure 8.6 as it draws revenue and cost data from all the other models. Showing all these linkages would also make this figure quite messy.

Scenario Projections

Key variables are differentially affected by the four scenarios, as detailed below. To project the outcomes for each scenario, initial conditions for each scenario were adjusted to achieve a zero Net Present Value (NPV) of any resulting budget surplus or deficit. Rather than fixing initial tuition and its rate of growth, we could have adjusted tuition each year to create zero deficits. However, this would make it difficult to compare scenarios.

Clash of Titans. In this scenario of business as usual on steroids, tuition grows steadily by 5% annually. Endowment grows steadily by an aggressive 8%. Percent tenure track is 80% to increase brand value. Percent tenured after the sixth year is 50%. The goal is to retain only the most productive faculty members. The undergraduate population grows slowly at 2% while the graduate population grows steadily at 6%. Administrative costs grow steadily at 6% as they have in recent years.

Hot, Flat, & Crowded. With competition among global universities intensifying, graduate enrollment decreases by 4% annually reflecting foreign students making different enrollment choices than in the past. Fewer graduate students result in a reduction of tenure track faculty to 30%. Tuition growth is limited to 2% and endowment growth slows to 4%. Growth of administrative costs is reduced to 3%.

Lifespan Mecca. Enrollment of older students seeking career changes or pursuing retirement interests results in the graduate population growing at 6% per year. The undergraduate population

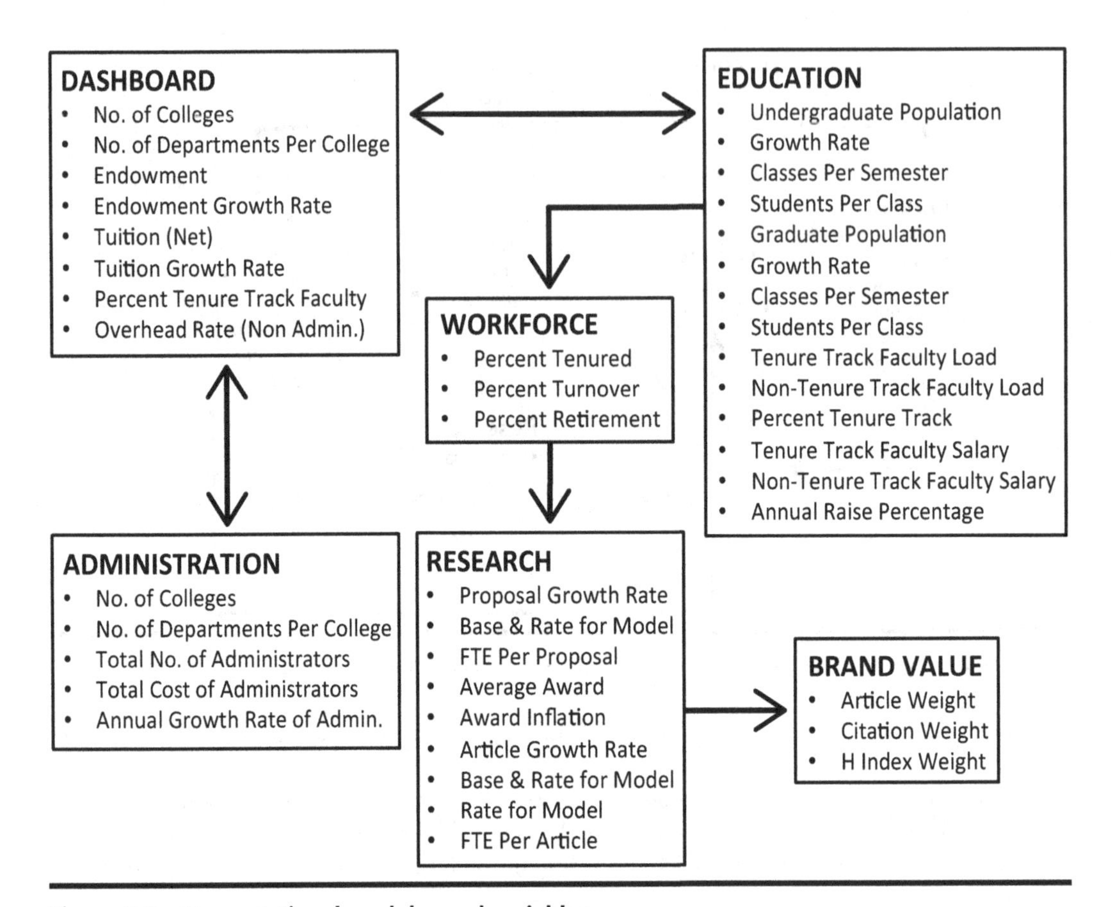

Figure 8.6 Computational modules and variables.

grows more slowly at 2%. Tuition increases are limited to 2% as much of this growth comes from people who are unwilling to pay constantly escalating tuitions. The percent tenure track faculty decreases to 30% because the MS and perhaps MA degrees being sought require more teaching faculty. Endowment grows slowly at 4%. Growth of administrative costs is limited to 3%.

Network U. Increased online offerings result in the graduate population growing quickly at 10% annually, while the undergraduate population grows more slowly at 2%. Classes become small discussion groups; class sizes vary from traditional numbers to much larger. The percent tenure track faculty decreases to 20% as the research enterprise becomes more focused on niches of excellence rather than trying to compete across the board. Tuition growth is necessarily limited to 1% in this highly competitive environment. The endowment grows very slowly at 2%, as most alumni have never set foot on campus. Administrative costs necessarily must decline by 5% annually.

Comparison of Projections

Figures 8.7–8.9 show the results of using these variable choices as inputs to the overall economic model of the university enterprise. Figure 8.7 portrays student population at year 20 and tuition for NPV equals zero. Note that the tuition does not differ greatly for each scenario. This is due to the model automatically adjusting the number of faculty members to meet demands. This is, of course, easier for non-tenure track faculty members than for those who are tenured.

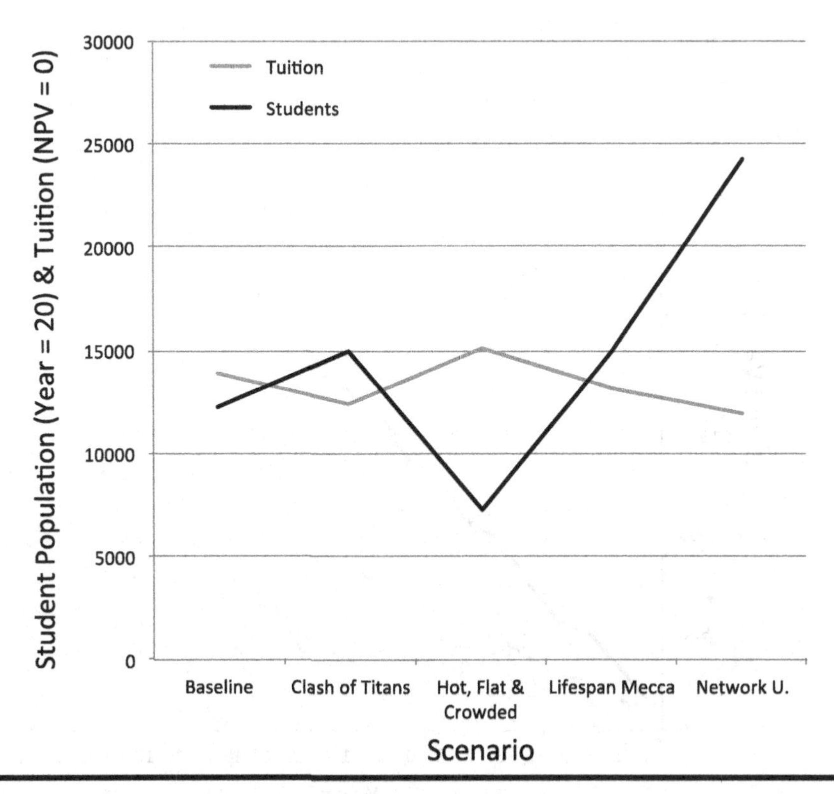

Figure 8.7 Student population (Year=20) and tuition (NPV=0).

The student population is depressed for Hot, Flat, & Crowded as graduate students choose to enroll at globally equivalent but less expensive universities. Lifespan Mecca attracts older American students that swell the graduate ranks. Not surprisingly, Network U leads to dramatic growth of online graduate students.

Figure 8.8 portrays the brand value for each scenario. Brand value for Clash of Titans dwarfs the other scenarios, the closest being the baseline. The other three scenarios drive needs to move away from emphases on graduate research conducted on campus. I have found that faculty members often have great difficulty thinking about such alternatives.

We would expect that technology-enabled Network U to have large classes of remotely connected students, probably very large for lectures and smaller for discussion sections. However, even the discussion classes are likely to be much larger than traditional campus classes. Figure 8.9 shows tuition versus class size in terms of numbers of times larger than the baseline.

The impact is fairly dramatic. As class sizes increase, the overall model automatically reduces the number of faculty members, which consequently substantially reduces costs. A rapidly growing student body (see Figure 8.7) while costs of delivery are plummeting enables cutting tuition from US$12,000 per semester to US$2,000.

Thus, an undergraduate degree would cost US$16,000 in total, assuming it requires eight semesters to earn enough credits to graduate. Of course, by this point the notion of semesters may be completely obsolete. Pricing will probably be by the course. How courses are bundled will be up to each student. Alternatively, pricing might be by the module, with students mixing and matching the modules to gain the knowledge and skills they seek.

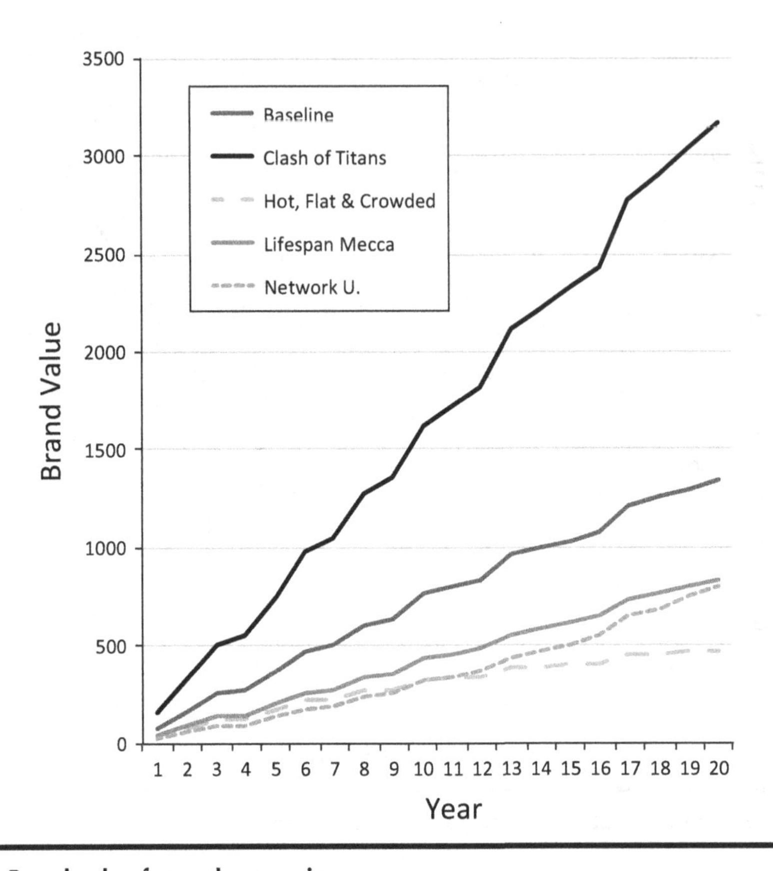

Figure 8.8 Brand value for each scenario.

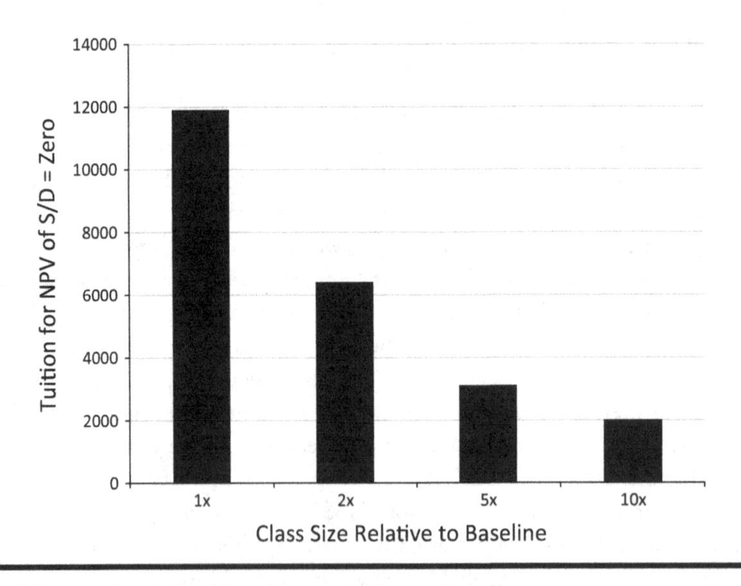

Figure 8.9 Tuition (NPV=0) for class sizes relative to baseline.

These tuition numbers may seem ridiculously low. However, Georgia Tech's online MS degrees in computer science and data analytics have tuitions of US$10,000 for the whole degree. They have 10,000 students enrolled and generate a tidy surplus. It helped greatly that AT&T and Accenture invested several million dollars to create these offerings.

This scenario easily causes one to consider what the university should do with its sizable investment in bricks and mortar. One possibility is that this infrastructure mainly serves the resident undergraduate student population, while the graduate population needs limited numbers of traditional classrooms and, of course, no dormitories and dining halls.

An overall comparison of these scenarios is as follows:

- Baseline: Keeps revenues and costs balanced across years with modest brand value; lower than Clash of Titans but higher than the other three scenarios
- Clash of Titans: Begins with slight deficit and then generates growing surplus as student population grows; brand value is strong due to high percent of tenure track faculty
- Hot, Flat, & Crowded: Leads to declining graduate enrollments and, in later years, steadily increasing deficits; brand value plummets
- Lifespan Mecca: Leads to strong growth of graduate enrollments and essentially zero deficits; brand value increases, relative to Hot, Flat, & Crowded, because of more faculty members being needed to serve increased enrollments
- Network U: Leads to exploding graduate enrollments; increasing class sizes, enabled by technology, dramatically lowers costs; brand value steadily decreases as larger class sizes lead to reduction of faculty size; initial deficits are replaced in later years by huge surpluses

Policy Implications

What are the strategic and tactical policy implications of the results found for this set of scenarios? These implications can be considered at two levels – across scenarios and within scenarios. The within scenario challenges and opportunities are likely to influence how across scenario issues are best addressed.

Across Scenarios. My experience has been that research universities most readily relate to Clash of Titans. If strong growth of the graduate student population is accompanied with large, yet acceptable, tuition growth as well as endowment growth, the above results show that pursuit of this scenario is fully viable. Problems will arise, however, if the other scenarios become salient.

The loss of graduate revenues portended by Hot, Flat, & Crowded could be quite difficult to sustain. I recently asked a Provost, "What would be the consequences if, for some political or economic reasons, all the Chinese graduate students disappeared?" He responded that, "We would lose $35 million in tuition revenues."

I asked, "What is Plan B?" He said, "Well, all research universities would be in this situation." I replied, "That does not sound like a Plan B." To be fair, the university in question is working to diversify its foreign graduate student population, which would decrease the chances of a single point of failure. Better yet would be approaches to encouraging more American graduate students, but this would require investing scarce resources in much larger stipends.

Lifespan Mecca could help this situation substantially, as the results shown above indicate. Professional educational offerings, supported by students' employers, plus affordable lifelong learning offerings could replace the graduate students lost to Hot, Flat, & Crowded. However, they would not be research students. Nevertheless, developing these offerings would be good hedges against the downside of Hot, Flat, & Crowded.

Network U could change the whole fabric of the university. "Guides on the Side" would replace many or most of the "Sages on the Stage." Flipped classrooms would become the norm. Online enrollment would soar. Faculty members' interpersonal skills would become core competencies. Students' sense of affiliation would mainly relate to the value being provided by education. The value of campus amenities and athletics would diminish.

One might embrace this scenario or simply choose to hedge against it. This would, at least, involve investing in capabilities to provide high value online offerings. One hybrid possibility would be investing in these capabilities to provide the Lifespan Mecca offerings noted above. This would provide the competencies, as well as some infrastructure, to enable scaling up when Network U becomes increasingly prevalent.

The bottom line is that one cannot just choose one of the scenarios. All of them must be addressed if only to define early warning signals of their emergence. More strategically, investments in Lifespan Mecca and Network U constitute hedges against Hot, Flat, & Crowded. A balanced investment portfolio across all scenarios is likely to be the best approach. It will mean that one cannot put all the eggs in the Clash of Titans basket, as that could be quite risky.

Within Scenarios. Clash of Titans presents a particularly difficult challenge. The current success model at most research universities requires faculty members to work harder and harder to achieve less and less success. Universities need to broaden their views of "gold standard" sponsors beyond National Institutes of Health (NIH) and National Science Foundation (NSF) to include other first-rate sponsors such as the National Aeronautics and Space Administration and the Office of Naval Research. Private foundations and industry sponsorship should be increased.

Universities also need to broaden their views of "gold standard" journals beyond current "A" journals. They should emphasize citations rather than impact factors, which have been shown to be irrelevant. A paper that earns 100+ citations in a low impact factor journal should be seen as a home run, not something to be dismissed.

University Presidents, Provosts, Deans, and Promotion & Tenure Committees need to communicate these changes to their faculties, particular junior faculty members. If everyone continues to pursue the old success model, there will be a lot less success, leading to pervasive frustration of junior faculty and much waste of human and financial resources. The outsourcing of evaluations of junior faculty needs to be tempered by more internal assessments.

Many universities have envisaged keeping Hot, Flat, & Crowded at bay by creating global campuses, the idea being that those who eschew matriculating in America can earn the same credential in Dubai or Singapore. There are merits to this idea, but also limits. I have experienced many faculty members of foreign extraction advocating the launch of a new campus in their native country. Campus leadership has encouraged this to the extent that the talent on the home campus was often diluted. Having a branch of CMU or MIT in every country is inevitably unsustainable, particularly in terms of brand value and quality of education.

On the other hand, making a Network U version of CMU and MIT globally accessible makes much more sense. An interesting hybrid involves pursuing a year or two online and the rest of the degree on campus. The key is for the university to make the investment to assure high quality online offerings that lead to the advertised knowledge and skills. This is not simply a matter of putting one's PowerPoint slides on the web. Proactive engagement of students in the learning experience requires that educators design this experience, monitor its evolutions and constantly improve it.

Lifespan Mecca requires careful attention to what students, ranging from mid-career professionals to eager-to-learn retirees, want and need to gain to achieve their educational aspirations, for example, promotions, new jobs, or simple mastery of history, music, or political science. Many traditional faculty members do not like to teach professionals and see history, music, and political science as "service courses." Success in this arena, therefore, may mean many fewer traditional faculty members.

Summary. Developing a strategy for addressing each scenario is necessary before one ponders how investments in these strategies might be leveraged across scenarios. The key point is that one does not know what mixture of these scenarios will emerge over the next 10–20 years. One's strategy across these scenarios needs to leverage opportunities while also hedging the downsides associated with these futures.

Impacts of Resource Demographics

To what extent do the above projections depend on the resources an institution has to respond to the forces of change? The economic model of research universities was extended to address this question (Rouse, Lombardi, & Craig, 2018). There are three forces of particular interest. They may work independently, but also may have combined effects on projected results:

- S1: Competition for federal dollars and publication in top journals is steadily increasing. The current success model at most research universities requires faculty members to work harder and harder to achieve less and less success, with proposal writing consuming increasing time and publication preparation receiving decreasing attention.
- S2: Foreign student applications to graduate programs have decreased in recent years due to competition from other countries and, more recently, concerns about US immigration policies. These professional master's degrees are typically "cash cows" for research universities, subsidizing many other aspects of the enterprise.
- S3: Highly polished, well-done Massive Open Online Courses (MOOCs) will increasingly succeed. Once the credentials associated with success in these online courses are acceptable to employers, it is easy to imagine a massive shift away from traditional classrooms for some categories of students, especially those seeking professional credentials and master's degrees where distance learning is already recognized and increasingly common.

We have extended and used the computational model to explore the implications of these forces for four specific research universities, two public and two private institutions, two large and two small. Size refers to resources available, not the number of students. The two large universities have immense endowments and very strong federal research support.

Well-resourced universities, such as the most successful among the top 100–200, will likely cope in different ways. Institutions that almost totally depend on tuition dollars, which typically fall outside the top group, will struggle to keep tuition competitive while avoiding large deficits.

Using 2016 data from the Center for Measuring University Performance (MUP) (Lombardi et al., 2000–2016), the model was fit to each of the four universities. We have not divulged the identity of each institution, but the model was explicitly fit to these particular universities. Fitting the model to specific institutions was not attempted before this paper, in part due to not having the MUP data.

Common assumptions across all institutions included undergraduate population growth rate of 3%, undergraduate tuition growth rate of 3%, graduate population growth rate of 4%, graduate tuition growth rate of 5%, endowment growth rate of 6.5%, endowment earnings of 5%, and discount rate of 4%. Sensitivity analyses showed that the overall results are not very sensitive to these assumptions in terms of plus or minus 1%. Negative values, in contrast, have a much larger impact.

Model Projections. The three scenarios are succinctly defined as follows: 1) S1: Status Quo, 2) S2: Graduate Student Population Declines by 5% Annually and, 3) S3: Graduate Tuition Declines to US$10,000 Due to Online Offerings. The results for these three scenarios are summarized in Figures 8.10 and 8.11. Note that class size is varied – to 10× or 1,000 – for the three instances of S3 rather than adding a fourth and fifth scenario. This reflects that fact that the external competitive driver is the same in all three cases. What differs is the institution's response to the scenario.

S3:$10K is the worst scenario, resulting in negative NPV (S/D) for everyone, because the number of students does not decrease while revenue decreases substantially. Three of the cases – S2, S3:10×, and S3:1K – lead to substantially reduced numbers of faculty, which undermines institutional publishing productivity and, hence, brand value. S3:1K is the most profitable because the number of students does not decrease but faculty numbers are cut by over 90%. Brand value, of course, plummets but only in a relative manner.

Institutions with significant resources are simply not going to let these futures happen to them. As discussed below, high-resource institutions have been the "first movers" in enabling S3:$10K. Thus, they are cannibalizing their professional master's "cash cows" before others do. They are likely to become the infrastructure platforms for others' educational content. They may also be

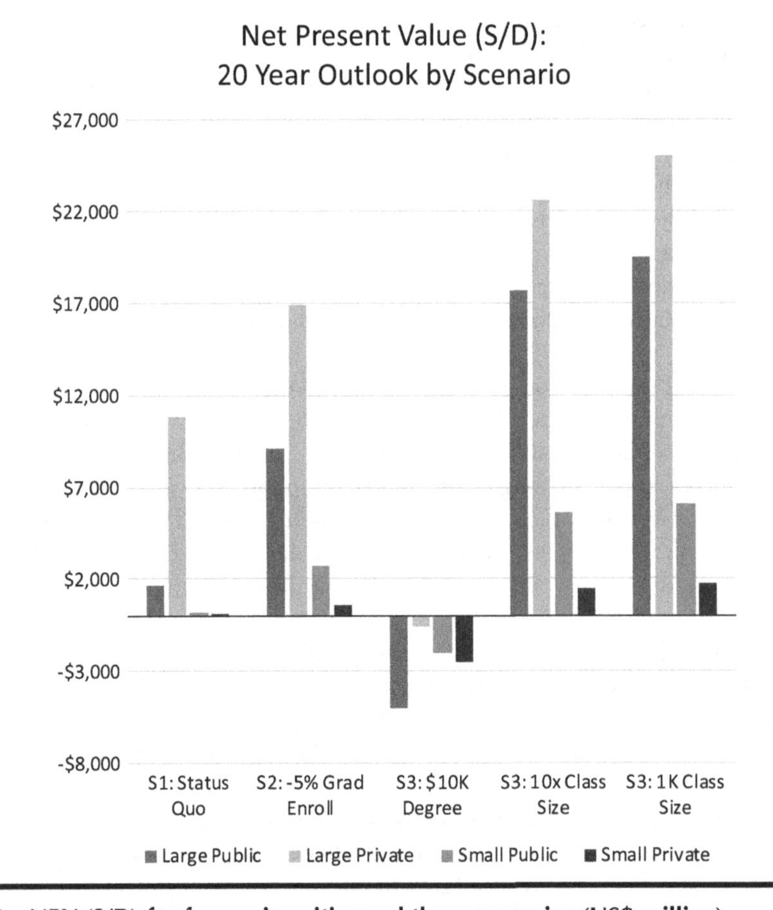

Figure 8.10 NPV (S/D) for four universities and three scenarios (US$ million).

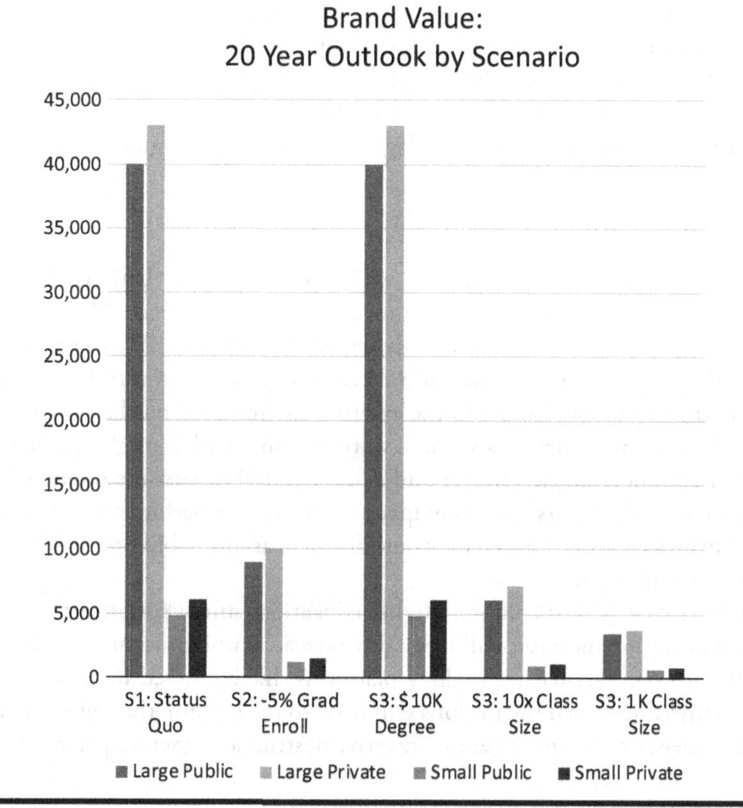

Figure 8.11 Brand value at year 20 for four universities and three scenarios.

content providers to resource poor institutions. This raises the possibility that these resource poor institutions will disappear or be absorbed by others (Azziz et al., 2017).

It is of particular note that the top institutions are driving all variations of scenario 3, with Coursera, edX, and Udacity being prime examples. These institutions have the resources to enable large experiments. In addition, they can attract major commitments from industry to underwrite these experiments and seed enrollments. AT&T and Accenture's large commitments to Georgia Tech for high-quality MS degrees in computer science and data analytics mentioned earlier illustrate how US$10,000 MS degrees can be possible. Lower-level players, where tuitions from professional graduate degrees are their only "cash cow," are at substantial risk.

Assuming class sizes of 1,000 raises the prospect of there not being enough students to fill these classes. However, US$10,000 professional MS degrees are likely to spur dramatic increases in demand, in part because this price point will easily fit within many large corporations' education budgets (Goodman et al., 2016). Nevertheless, the higher brand value institutions may dominate this market, to the significant detriment of the lower brand value institutions.

Comparing Institutions. How do the different scenarios affect the four institutions studied? Brand value decreases due to diminishing returns from research sponsors affects all institutions similarly. The ratios of brand value of large institutions to small institutions range from 4.7 to 7.7 across the scenarios. Thus, the top-ranked institutions will likely remain on top. The substantially declining research productivity of all four institutions should be a major concern in terms of economic development, national security, etc.

The change of NPV differs significantly across large and small institutions, particularly for S3. The two large institutions average NPV=-US$2.8 billion, while the two small institutions average NPV=-US$2.3 billion. The year 20 revenues for the large institutions average US6.8 billion, while the two small institutions average US$1.0 billion. Clearly, the small institutions are not in a position to weather such losses due to technology transforming their graduate education business.

Summary

It is unknown what mix of these scenarios will actually emerge. Universities need strategies and investments that enable robust responses to whatever mix emerges. Models such as the one extended and exemplified here, and more fully explored in (Rouse, 2016), provide institutional leaders a method of exploring the impacts of various policy decisions within their institutions, as well as assessing the impact of changes in the external environment on their institution.

Predictions for the three scenarios serve as warnings about what might happen if universities persist with their current strategies. Meyer and Zucker (1989) discuss the notion of "permanently failing organizations," where persistence compensates for lack of performance. Muddling through will not work when faced with the scenarios outlined here as losing billions of dollars is not a realistic option for many institutions.

Fundamental change is in the offing. Higher education cannot sustain its current cost structures. The limits of tuition increases will inevitably be reached, significantly facilitated by increasingly powerful and sophisticated technology platforms, likely offered by institutions with high brand values. Many educational institutions will need to reconfigure their operations, restructure their financial models, or disappear amidst "creative destruction" (Schumpeter, 1942).

Energy and Transportation

Transportation is the segment of our economy that consumes the most energy. When propelled by fossil fuels, vehicles yield the most CO_2 emissions. In this section, I consider two technology innovations, which we explored using systems dynamics models to power the games hosted in the *Immersion Lab.*

Battery Electric Vehicles

Substantial environmental and energy challenges are driving the pursuit of alternative powertrain technologies, which nominally include engine, transmission, driveshafts, differentials, and the final drive. Emerging alternative fuel vehicles are showing their potential to address these challenges. However, diffusion of new technologies has many complications. Working with General Motors, we investigated the impacts of individual and organizational parameters on the adoption of battery electric vehicles (BEVs).

We employed system dynamics modeling to create the representation in Figure 8.12. Mathematical relationships among different variables were derived. The impacts of government rebates, manufacturer willingness to invest, and consumer purchasing preferences on economic and environmental issues were addressed using scenario analysis.

Three major stakeholders in the California automobile market were considered (government, manufacturer, and consumer). The types of powertrain systems considered included small/midsize internal combustion engines (ICE), large size ICE, hybrid, EV, and fuel cell electric vehicles.

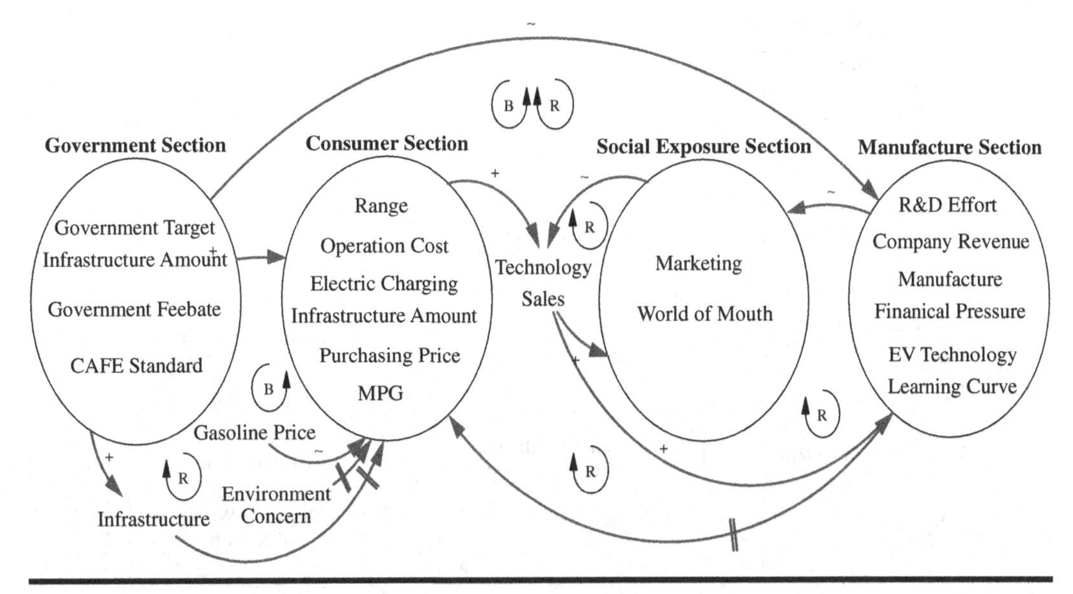

Figure 8.12 Systems dynamics model of BEV adoption (Liu, Rouse, & Hanawalt, 2018).

Near-term impacts of government rebates, both federal and state, were found to be important to launch the market.

However, the model suggested that long-term impacts will come primarily from product familiarity, consumer preference, and technology competitiveness. This supports the importance of investments in R&D and advertising. Such investments could be augmented by government support of manufacturers or related research organizations. Rather than depending on short-term rebates to consumers, fundamental improvements of technology and infrastructure, e.g., charging stations are more resilient ways to achieve a new technology's long-term self-growth.

BEVs were found to be significantly more environmentally friendly if the electricity used to charge the vehicles was not produced by coal-fired electric plants. Green electricity generation will lead to larger and more stable environmental improvements in the long term. Furthermore, pure green electricity production affects CO_2 emissions beyond just the vehicles. However, totally switching to green energy production in a short time is highly unlikely. Nevertheless, the model suggests the importance and value of paying more attention to changing production to green energy methods.

Autonomous Vehicles

The transportation industry is facing a revolution similar to when machines replaced animals one century ago. Humans may for the first time be fully out of the control loop of personal transportation. However, this revolution involves considerable disruption and uncertainty. Nevertheless, autonomous vehicles (AVs) will increasingly impact the automobile market.

We mapped various causal relationships during this significant transition to understand the impacts of different phenomena. The systems dynamic model in Figure 8.13 was constructed including two different transportation methods (personal owned vehicle and car services) and three autonomy levels (non, semi, and fully).

Consumer choices, product familiarity, and acceptance were modeled to represent purchasing behavior. The US auto insurance industry is likely to be substantially impacted by AVs. Vehicle

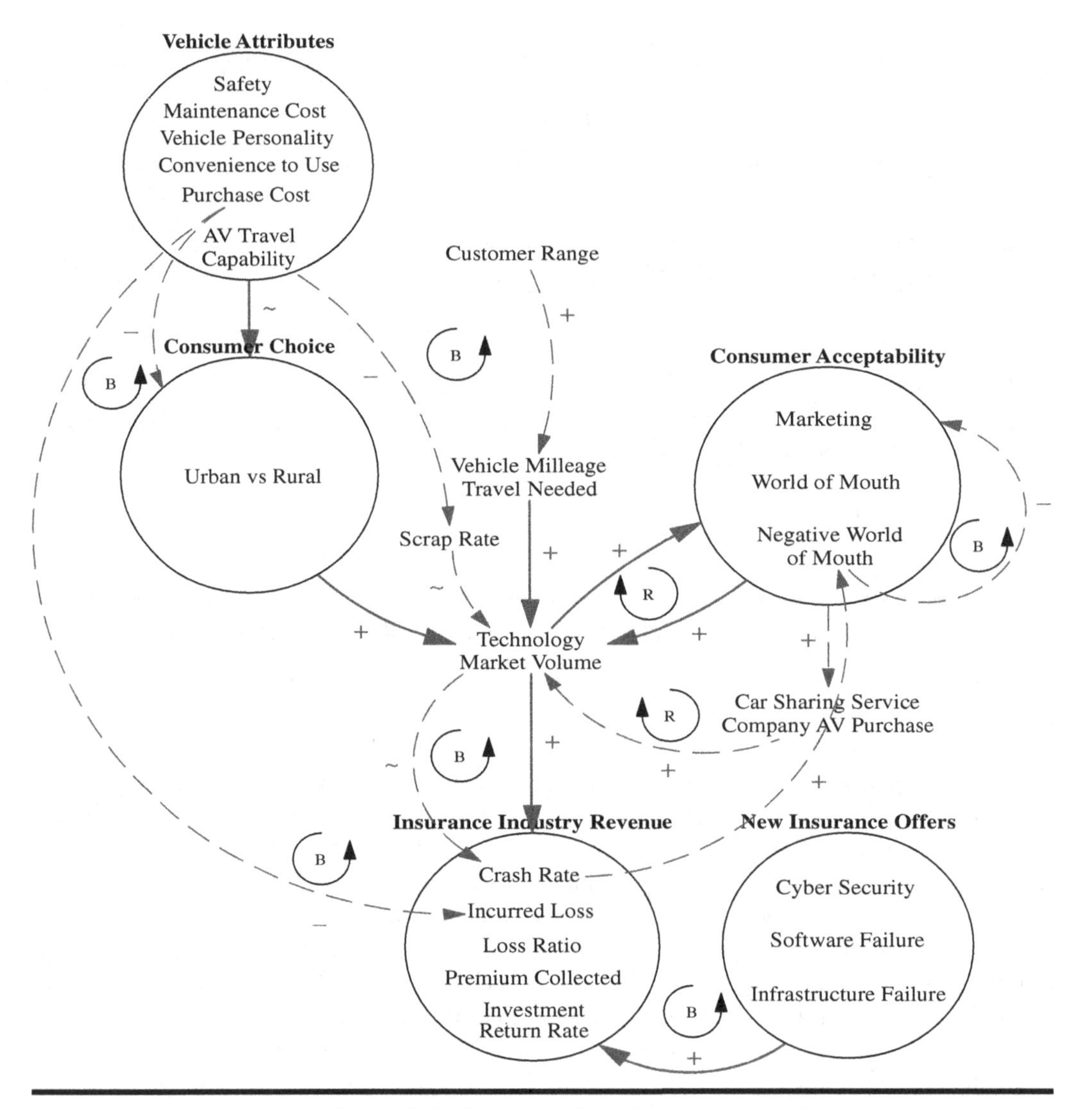

Figure 8.13 Systems dynamics model of AV adoption (Liu, Rouse, & Belanger, 2020).

crash rates and loss ratios were considered to calculate the insurance industry's premium collections. Different scenarios were quantified and discussed with key stakeholders. Several important causal loops were identified that will help achieve the faster growth of the technology.

With gradually improving vehicle driving assistance technologies, AVs are expected to debut in this decade. As a revolutionary way of personal transportation, it is very promising from various perspectives, including enriching personal mobility, reducing energy consumption, and dramatically decreasing vehicle accidents. Yet, not everyone will see economic gains.

Working with Accenture, a primary motivation for this effort was to understand the impacts of AVs on their insurance industry clients. Every state in the US has regulations that limit auto insurance premiums to the costs of insurance claims. Thus, insurance companies do not make profits on premiums. They make money by investing the premium monies until these funds are needed to pay claims.

The model predicts that insurance industry premiums collected will continue growing until the penetration rate of AVs becomes significant. At that point, the frequency of accidents and hence claims will decrease. Once the AV technology takes off, the industry premiums collected will be dramatically reduced.

Another consideration is the likelihood that people will use car services that own AVs rather than own the vehicles themselves. As these vehicles will be highly utilized, the total number of vehicles on the road will be reduced. Thus, the economic scale of the insurance industry will be further reduced, which will lead to decreased premiums collected. Reduced accidents and fewer cars on the road combine to result in substantial reductions of insurance premiums collected.

Summary

The models in Figures 8.12 and 8.13 were used to drive the displays in the *Immersion Lab* to pursue a range of "What if?" scenarios with key stakeholders. Intense discussions focused on key assumptions, and exploring best case and worst case scenarios. These explorations amounted to gaming with the environment and automotive markets.

From an energy and climate perspective, BEVs and AVs, which will also be electric, will reduce emissions per vehicle and reduce the number of vehicles, contingent on moving away from coal-fired electricity power plants. Beyond positive environmental effects, these automotive innovations will also have substantial additional benefits for the people who will benefit from these vehicles, e.g., disabled and older adults whose mobility will be greatly enhanced by AVs.

Comparison of Games

The computational games discussed in this chapter were hosted in the *Immersion Lab*. All of these sessions prompted serious stakeholder involvement. A common comment by one executive or another, typically to a colleague, was, "Did you realize that X affects Y?" There was substantial mutual discovery.

Table 8.1 compares the three domains addressed in this chapter. The objectives, stakeholders, computational models, data sources, and findings differ substantially for these games. Yet, all but one of the four were hosted in the same *Immersion Lab.*

They all involved multi-disciplinary teams pursuing "What if?" questions of personal importance to them. Their discussions and debates were wide-ranging, checking assumptions, unearthing relationships, positing policies, and assessing responses.

Conclusions

Serious games can be interesting and fun. Certainly, the players of these games reported enjoying themselves. However, these games were "serious" in the sense that they addressed important real-world problems for which the players had responsibilities to address. They did not just want to "win the game." They wanted insights and ideas they might actually implement with evidence-based confidence.

Serious games provide venues for exploration and experimentation. Hosted in an *Immersion Lab*, they provide venues for collaboration, laced with discussions and debates. There are often

Table 8.1 Comparison of Domains

Attributes	*Domain*		
	Healthcare	*Education*	*Energy*
Stakeholders	Provider executives & consultants	Senior institutional leadership	Automobile & insurance companies
Underlying representation	Agent-based models	Economic models in MS Excel	Systems dynamics models
Scope of representation	66 agents in NYC; 700,000 in DC	Individual institutions	California vehicle market
Central predictions	M&A in NYC; Overdoses & deaths in DC	Net Present Value of surplus/deficit	Market penetration & environmental impacts
Important findings	Best M&A matches; Impacts of social interventions	Resource base of institutions affect strategy options	Automakers need to invest; Insurance companies will struggle

disagreements. These are addressed in the game, not across the table or in the hallway. Rather than argue about an idea, players try it out in the game. This creates evidence – or the lack of evidence – for why an idea is good or bad.

Harking back to Chapter 4, typical interactions in the *Immersion Lab* enhance collaborators' team mental models. They gain understanding of other stakeholders' values, concerns, and perceptions. They come to understand each other's strengths and weaknesses, in terms of disciplinary backgrounds, organizational priorities, and political leanings.

Such serious games can be wonderful border-crossing mechanisms. The borders crossed are between disciplines, organizations, and even belief systems. Once players are able to see issues from both sides on these borders, amazing creativity often emerges. People envision policy options, for example, that they never would have previously imagined.

Serious games are a compelling example of the power of games. People can explore these worlds to identify new approaches to serious problems, and marshal the evidence to support these approaches. My experience has been that stakeholders leave the game playing experience with greatly enhanced enthusiasm and confidence.

References

Azziz, R., et al. (2017). *Mergers in Higher Education: A Proactive Strategy to a Better Future?* New York: TIAA Institute.

CSGNET (2007). *Data Sources: The Rise of "Older" Graduate Students.* http:// www.cgsnet.org/portals/0/pdf/DataSources_2007_12.pdf.

Fahey, L., & Randall, R.M. (Eds.). (1998). *Learning From The Future: Competitive Foresight Scenarios.* New York: Wiley.

Friedman, T.L. (2005). *The World Is Flat.* New York: Farrar, Straus and Giroux.

Friedman, T.L. (2008). *Hot, Flat, and Crowded.* New York: Farrar, Straus and Giroux.

Goodman, J., Melkers, J., & Pallais, A. (2016). *Can Online Delivery Increase Access to Education? Working Paper No.* W22754. Cambridge, MA: National Bureau of Economic Research.

Kamenetz, A. (2009). *Who Needs Harvard? Fast Company*, September.
Liu, C. Rouse, W.B., & Belanger, D. (2020). Understanding risks and opportunities of autonomous vehicle technology adoption through systems dynamic scenario modeling – The American insurance industry. *IEEE Systems Journal*, 14 (1), 1365–1374. https://doi.org/ 10.1109/JSYST.2019. 2913647.
Liu, C., Rouse, W.B., & Hanawalt, E. (2018). Adoption of powertrain technologies in automobiles: A system dynamics model of technology diffusion in the American market. *IEEE Transactions on Vehicular Technology*, 67 (7), 5621–5634.
Lombardi, J.V., et al., (2000–2016). *The Top American Research Universities: Annual Report, 2000–2016.* Arizona State University and University of Massachusetts Amherst. http://mup.asu.edu.
Meyer, M.W., & Zucker, L.G. (1989). *Permanently Failing Organizations*. Washington, DC: Sage Publications.
North, G. (2009). *MIT Calls Academia's Bluff.* www.LewRockwell.com.
Rouse, W.B. (2016). *Universities as Complex Enterprises: How Academia Works, Why It Works These Ways, and Where The University Enterprise Is Headed.* New York: Wiley.
Rouse, W.B., Lombardi, J.V, & Craig, D.D. (2018). Modeling research universities: Predicting probable futures of public vs. private and large vs. small research universities. *Proceedings of the National Academy of Sciences*, 115 (50), 12582–12589.
Schoemaker, P.J.H. (1995). Scenario planning: A tool for strategic thinking. *Sloan Management Review*, Winter, 25–40.
Schumpeter, J. (1942). *Capitalism, Socialism and Democracy*. New York: Harper & Brothers.
Schwartz, P. (1991). *The Art of the Long View: Planning for the Future in an Uncertain World.* New York: Currency Doubleday.
Tolk, A., Rouse, W.B., Pires, B.S., Cline, J.C, Diallo, S.Y., & Russell, S.A. (2023). Applicability of artificial societies to evaluate health care policies. *Simulation in Healthcare*. https://doi.org/10.1097/SIH .0000000000000718.
Yu, Z., Rouse, W.B., Serban, N., & Veral, E. (2016). A data-rich agent-based decision support model for hospital consolidation. *Journal of Enterprise Transformation*, 6 (3/4), 136–161.

Chapter 9

Game Industry

Introduction

This chapter outlines the nature of the game industry, broadly defined. Its relationships with the film industry and theme parks are considered. Earlier reviews include Castronova (2006), Bogost (2010), and Polfeldt (2020). These excellent and insightful books are narrower than this book in that they do not consider training simulators or sports in general, although their insights into mass market video games are valuable.

Producers of mass market games are highlighted, e.g., Activision, Electronic Arts, Nintendo, et al. I describe the popular game stations, including PlayStation, Xbox, and Switch, as well as popular headsets for augmented and virtual reality (AR/VR). Bespoke games are not mass market offerings and include, for example, training simulator companies, RAND for Pentagon board games, and Army Games Studio for all of DoD.

I address game development methods and tools in terms of both methodologies and game engines. A human-centered design methodology is elaborated and illustrated as it applies to serious games. Game engines enable efficient game development with assorted reusable assets. The most popular game engines are compared, with an emphasis on Unity and Unreal.

This chapter concludes with a characterization of the rapidly evolving competitive landscape in terms of market forces and technology trends. Game offerings are influencing the film industry and theme parks. The apparent strategies of the major players are rapidly evolving. I summarize and project these trends.

Companies

Mass Market Games

Producers of mass market games include Activision, Electronic Arts, Microsoft, Nintendo, and Sony. The top ten mass market games producers in terms of game-related revenues are listed in Table 9.1. Industry revenues were US$60 billion in 2021. In contrast, global board game revenue amounted to US$15 billion, with younger players being the major consumers.

Compared to the revenues noted in Table 9.1, the total revenues of companies such as Apple, Google, and Microsoft are enormous. Apple generated US$394 billion of revenue in 2022, Google

DOI: 10.4324/9781003491927-9

Table 9.1 Leading Companies and Best Selling Games

Company	*Annual Games Revenue (US$B)*	*Best Selling Game*
Tencent	7.6	*Honor of Kings*
Sony	4.4	*Call of Duty* (PlayStation)
Apple	3.7	*Apple Arcade*
Microsoft	3.2	*Call of Duty* (Xbox)
NetEase	2.7	*Identity V*
Google	2.4	Mine*craft*
Activision	2.3	*Overwatch*
Electronic Arts	1.9	*FIFA*
Nintendo	1.3	*Mario Kart 8* (Switch)
Take-Two Interactive	1.3	*Grand Theft Auto V*

US$280 billion, and Microsoft US$198 billion. It is notable, however, that a significant portion of these larger revenues is for devices and infrastructure on which games can be played.

ESA (2022) reports there are 216 million active game players in the US, 76% over the age of 18. Time spent playing is over 7 hours per week (53%), over 3 hours per week (78%), or limited to 1–3 hours per week (22%). Players report that game playing improves their cognitive skills (88%), creative skills (86%), teamwork and collaboration skills (86%), communication skills (63%), and leadership skills (61%).

Bespoke Games

Bespoke games are developed especially for a particular user cohort that is not commercially available for the general public to buy. All the games in Chapter 8, as well as many in Chapter 7, are bespoke. The range of investors and customers for these games is highly diverse, making it difficult to estimate revenues.

Producers of bespoke games include training simulators, with top companies being CAE, L3Harris, Thales, and Saab. Training simulators generated US$26 billion of revenue in 2022. RAND's board games for the Pentagon (Bauman, 2020) as well as the games produced by the Army Games Studio for all of DoD, e.g., *America's Army* (Skelley, 2021) are also bespoke. It is important to note that bespoke games are almost always targeted at specific problems rather than mass entertainment.

Sports and Gambling

Professional sports generate US$512 billion of revenue globally. College sports generate US$16 billion for Division I in the US. Legal gambling generates US$60 billion in the US, much of it related to sports betting. Thus, these types of games generate roughly ten times the revenue of mass market games. As discussed in Chapters 5 and 6, this ecosystem is laced with money and corruption, so actual revenues are likely much greater.

Related Products

Game Stations

Twenty-eight percent of games are played on game stations rather than personal computers, tablets, and smartphones. Units sold for PlayStation, Xbox, and Switch are summarized in Table 9.2. Note in Table 9.1 that PlayStation and Xbox both offer *Call of Duty*, a video game series and media franchise published by Activision, starting in 2003.

AR/VR Headsets

Augmented reality and virtual reality refer to computer-generated simulations that integrate the real world (AR) or are entirely self-contained (VR). Units sold for Quest, Playstation VR, and Vision Pro are listed in Table 9.3. Facebook, now Meta, acquired market leader Oculus in 2014 for US$2 billion. Apple's Vision Pro was introduced in June 2023. Google, Nintendo, and Samsung have discontinued their AR/VR offerings.

Game Development Methods and Tools

Game design and development begins with learning objectives as discussed in earlier chapters. This leads to the conceptual design of a game and, once proven, a detailed design. A systematic design methodology yields better results. Design is followed by game development. For online games, a range of games engines is available.

Methodology

Fullerton (2008) provides game design guidance with many vignettes. Her recommended design process is quite reasonable, perhaps what a systems engineer would expect. The game documentation resulting from such a design process can be substantial. For example, Hall (1992) provides the

Table 9.2 Leading Game Station Providers and Units Sold

Company	*Game Station*	*Units Sold (Millions)*
Sony	PlayStation	155
Nintendo	Switch	130
Microsoft	Xbox	48

Table 9.3 Leading AR/VR Headsets

Company	*Headset*	*Units Sold (Million)*
Meta/Oculus	Quest	20
Sony	PlayStation VR	5
Apple	Vision Pro	0.4

Doom Bible, a 49-page design document that summarizes the characters, story, actors, maps, and weapons for this classic winner of The Game Awards' Best Action Game of 2016 (Romero, 2023).

We have adopted a methodology termed human-centered design that addresses the values, concerns, and perceptions of all stakeholders in designing, developing, deploying, and employing policies, products, and services (Rouse, 2007, 2015). The basic premise is that the major stakeholders need to perceive policies, products, and services to be valid, acceptable, and viable.

Valid policies, products, and services demonstrably help solve the problems for which they are intended. Acceptable policies, products, and services solve problems in ways that stakeholders prefer. Viable policies, products, and services provide benefits that are worth the costs of use. Costs here include the efforts needed to learn and use policies, products, and services, not just the purchase price.

How can human-centered design be applied to game design? What does it mean for a game to be valid, acceptable, and viable? Consider these questions in the context of the serious games discussed in Chapter 8. Validity for the users of these games concerns the sources and accuracy of the data involved and the appropriateness of the computational models employed. Users will often pose test cases where they know what should happen. If their expectations are fulfilled, their perceptions of validity improve.

Acceptability for these users concerns the ways information is portrayed and how they can manipulate controls to operate the game. The ability to switch between tables and graphs can be highly valued. The touch screens in Figure 8.1 were popular as users could simply touch what they wanted rather than being forced to designate choices via keyboard. The very large screens in Figure 8.1 were helpful when there were many simultaneous users.

Viability for these users related to the benefits of playing the games versus the time required to learn and play effectively. Benefits include the insights gained from the games as well as the enhanced team mental models resulting from playing. We found that the availability of a highly knowledgeable human assistant greatly facilitated users moving up the game learning curve.

Beyond the use of the game, another consideration is the design of the game. Involvement of representative game players early in the design process results in better games. In an ongoing effort to create an online employee training game, we have begun by creating a board game with substantial user participation. It is much easier to prototype, evaluate, and revise board games. We will get to the game engine for the online game after users assess that our overall board game concept is sound.

Game Engines

A game engine is a software development framework with settings and configurations that optimize and simplify the development of video games across a variety of programming languages. The number of active end users of Unity, Unreal, and ARKit is listed in Table 9.4. Unity was

Table 9.4 Leading Game Engine Providers and Active End Users

Company	*Game Engine*	*Active End Users (Billions)*
Unity Technologies	Unity	3.9
Epic Games	Unreal Engine	2.7
Apple	ARKit	On 1.6 Billion iPhones

introduced in 2005 and Unreal in 1998. ARKit, a relative newcomer, was introduced by Apple in June 2017.

Table 9.5 compares the two leading games engines – Unity and Unreal. This compilation was adapted from Arora and Johns (2023).

They both represent good choices. I have found that the strongest determinant of which is chosen in the organizational experience base with either tool.

Competitive Forces

The evolution of *Zelda* sets the stage for considering the rapidly evolving competitive landscape in this industry. The *Legend of Zelda: Tears of the Kingdom* was released on May 12, 2023. Zelda is "an action game, it encourages experimentation, using things in new ways. It is highly interactive and graphical both horizontally and vertically" Park, 2023a). "An open-world game, one that tantalizes players to explore a vibrant environment full of ambitious quests and powerful equipment" (Small & Taylor, 2023).

> This video game franchise that started in 1986 with a pixelated map guarded by ghosts and goblins has evolved into an elaborate topography of mountain ridges, coastline villages and enemy hide-outs. The gameplay has also become more riveting, with puzzle-box designs and environmental storytelling that encourage exploration.
>
> **(Small & Taylor, 2023)**

Its predecessor sold 29 million copies, often purchased with the Nintendo game console (Park, 2023a).

> Since the turn of the century, Eiji Aonuma has been the franchise master of the *Zelda* series. He's worked on every title since 1998's *Ocarina of Time*, often called the *Citizen Kane* of video games thanks to innovations that inspired *Grand Theft Auto* and

Table 9.5 Unity versus Unreal Game Engines (Adapted from Arora & Johns, 2023)

Attribute	*Unity*	*Unreal*
Source Code	Read-only access	Full access
Release Date	2005	1998
Languages	C#	C++
Community	50% market share	13% market share
Documentation	Detailed	Detailed
Asset Store	65,000	16,000
Graphics	Excellent	Better
Rendering	Fast	Faster
Pricing	Free; pay for prof. version	Free; royalties on resales

countless other titles. *Tears of the Kingdom* released to universal critical acclaim and sold 10 million copies in its first three days. It's a near-impossible accomplishment to follow a masterpiece with another one—and he's not even done cooking up ideas for the series.

(Park, 2023b)

The Economist has published a series of reports on the rapidly evolving competitive game industry landscape that can be characterized in terms of market forces and technology trends, including the apparent strategies of the major players. The first report (Economist, 2023a) addressed the challenges faced by Disney.

> Many predicted that this surge of niche content would bring down mainstream hit-makers. They were mostly wrong. Infinite choice in entertainment has ruined the companies which produced middling content that people watched because there was nothing else on—witness the collapse in broadcast-television ratings. But those at the very top of the business have thrived. When anyone can watch anything, people flock to the best. Global streamers like Netflix and Amazon have more than 200 million direct subscribers, once an unimaginable number.
>
> Those who have fared best at a shrinking box often are the owners of IP (intellectual property) that is already popular. As people visit cinemas less often and competition intensifies, studios have pumped money into films that people will turn out to see even when they go only three or four times a year. It has not been a golden age for cinema, but for those at the top it has been a profitable one.
>
> Now technology is shaking things up again. Online distribution has enticed tech firms that make the hardware and software used for streaming. Hollywood initially wrote off the nerds. But the nerds have enough money to take creative risks. Last year Apple won the best-picture Oscar with *Coda*, a comedy-drama partly in sign language, less than three years after it entered the film business. The more new content these new producers make and sell below cost, the greater the risk that older studios will fall from the top tier of media into the perilous middle.
>
> Inventions like game engines, which help with the creation of virtual sets, are lowering barriers to entry. Generative artificial intelligence, which can already make rudimentary video, may eventually lower these barriers further. Perhaps the most dramatic way technology could disrupt the culture business is by creating new categories of entertainment. Young adults in rich countries already devote more time to gaming than to broadcast television.
>
> Microsoft's proposed acquisition of Activision-Blizzard, whose games include *Call of Duty* and *Candy Crush*, is worth nearly ten times what Amazon paid for Metro-Goldwyn-Mayer, home of James Bond and Rocky Balboa. Movies based on games are becoming as popular as games based on movies.

Games are also infiltrating theme parks.

> As gaming matures, it is not just rivalling other media (Economist, 2023b). Rather like a ravenous *Pac-Man*, it is gobbling them up. After music, gaming clips are the

> biggest content category on YouTube. When video games were just electronic toys, this might not have mattered. But as games expand and spill into other formats, it is becoming clear that whoever dominates gaming is going to wield clout in every form of communication. In every sense, the future of the media is in play.
>
> Making a blockbuster game is now like making a blockbuster movie (Economist, 2023c). As technology lets games grow larger and more lifelike, they have taken on Hollywood-style budgets and timetables. And as the line between film and digital games blurs, that has two effects. One is that labor markets and production techniques for gaming converge with those of the film business, to the point where some envisage a single production process. The other is that game studios become more focused even than film studios on monetizing a few successful franchises.
>
> If the lawyers don't intervene, unions might (Economist, 2023d). Studios diplomatically refer to AI assistants as "co-pilots", not replacements for humans. But workers are taking no chances. The Writers' Guild of America, whose members include game scriptwriters, said in March that plagiarism is a feature of the AI process. In Hollywood, it is threatening strikes. Upset creatives may press pause on the games business, too.

Nevertheless, generative AI is on on the horizon (Park et al., 2023). A research consortium led by Stanford University reports that

> Generative agents create believable simulacra of human behavior for interactive applications. In this work, we demonstrate generative agents by populating a sandbox environment, reminiscent of *The Sims*, with twenty-five agents. Users can observe and intervene as agents plan their days, share news, form relationships, and coordinate group activities.

The game scriptwriters' concerns seem well founded.

Yet, there are other upsides to these trends. Economist (2023e) reports that games are a weapon in the war on disinformation: They are effective at teaching players how to spot falsehoods online. As important as this trend is (Rouse, 2023), it may provide small comfort to the game scriptwriters.

Overall, it seems reasonable to conclude that strong competitive forces, driven and enabled by technology trends, will strongly affect how the games of the game industry are played. Games, films, and theme parks will be significantly affected. The really big players, such as Apple, Google, and Microsoft – with their aforementioned combined annual revenues of nearly US$1 trillion – have the resources and intentions to transform these industries.

Conclusions

The game industry includes those who provide mass market games, bespoke games (e.g., training simulators), professional and college sports, and gambling venues. The mass market portion of the games industry is roughly ten times the size of the film industry, but only one tenth the size of professional and college sports, which is only half the size of Apple, Google, and Microsoft. The scales of many of the players in this ecosystem are truly amazing.

Yet, the trends favor those adept at leveraging the technologies in creative ways to both provide new gaming experiences and dramatically lower the costs of competing. Realization of these objectives will disrupt business as usual. It may initially yield uneven quality. However, the future of gaming is very bright and pervasive. As Chapter 10 outlines, education, management, and operations will be transformed. New and emerging value propositions will be broadly beneficial, once we fully understand and adapt accordingly.

References

Arora, S.K., & Johns, R. (2023). *Unity* vs *Unreal*: Which Game Engine Should You Choose?. March 23, https://hackr.io/blog/unity-vs-unreal-engine.

Bauman, M. (2020). Now you can play RAND games at home. *The RAND Blog*. https://www.rand.org/blog/rand-review/2020/11/now-you-can-play-rand-games-at-home.html/.

Bogost, I. (2010). *Persuasive Games: The Expressive Power of Videogames*. Cambridge, MA: MIT Press.

Castronova, E. (2006). *Synthetic Worlds: The Business and Culture of Online Games*. Evanston, IL: University of Chicago Press.

Economist (2023a). Disney's troubles show how technology has changed the business of culture. *The Economist*, January 19.

Economist (2023b). As video games grow, they are eating the media. *The Economist*, March 23.

Economist (2023c). Moviemaking and gamemaking are converging. *The Economist*, March 23.

Economist (2023d). How AI could disrupt video-gaming. *The Economist*, April 5.

Economist (2023e). Games are a weapon in the war on disinformation: They are effective at teaching players how to spot falsehoods online. *The Economist*, April 5.

ESA (2022). *Essential Facts About the Video Game Industry*. Washington, DC: Entertainment Software Association.

Fullerton, T. (2008). *Game Design Workshop: A Play Centric Approach to Creating Innovative Games*. Amsterdam: Elsevier.

Hall, T. (1992). *Doom Bible*. Richardson, TX: id Software.

Park, G. (2023a). The new Zelda game is like nothing we've ever seen before. *Washington Post*, April 26.

Park, G. (2023b). The genius behind Zelda is at the peak of his power – and feeling his age. *Washington Post*, May 20.

Park, J.S, Obrien, J.C, Cai, C.J., Morris, M.R., Liang, P., & Bernstein, M.S. (2023). Generative agents: Interactive simulacra of human behavior. *arXiv*, April.

Polfeldt, D. (2020). *The Dream Architects: Adventures in the Video Game Industry*. New York: Grand Central Publishing.

Romero, J. (2023). *Doom Guy: Life in First Person*. New York: Abrams.

Rouse, W.B. (2007). *People and Organizations: Explorations of Human-Centered Design*. New York: Wiley.

Rouse, W.B. (2015). *Modeling and Visualization of Complex Systems and Enterprises: Explorations of Physical, Human, Economic, and Social Phenomena*. Hoboken, NJ: John Wiley.

Rouse, W.B. (2023). *Beyond Quick Fixes: Addressing the Complexity & Uncertainties of Contemporary Society*. Oxford, UK: Oxford University Press.

Skelley, K.D. (2021). Army Game Studio levels up soldier recruitment and training. July 14, https://www.army.mil/article/248435/army_game_studio_levels_up_soldier_recruitment_and_training.

Small, Z., & Taylor, R. (2023). How the Legend of Zelda changed the game. *The New York Times*, May 4.

Chapter 10

Future of Gaming

Introduction

There is no doubt that gaming has a future. Its rich history is, at the very least, a strong predictor of people playing, learning, competing, and achieving. However, to what extent can games become predominant in supporting the target populations to achieve the learning objectives discussed in Chapters 7 and 8? While this possibility makes great sense, there are traditional forces that will ardently support the status quo of education and training. This chapter discusses mechanisms for mitigating such opposition and gaining the benefits that gaming promises.

Possible Futures

Assume it is 1980 and you are trying to imagine 2030. The internet existed but hardly anybody knew about it. The golden age of the fax machine was in full tilt. Cell phones had just emerged. Digital cell phones were on the distant horizon. Analog technology still dominated. CGI in movies was just emerging, although Alfred Hitchcock experimented with it in *Vertigo* in 1958.

Actually, sitting here in 2024, it is difficult to imagine 2030. What should we expect from Chat GPT? What about the metaverse? I doubt we will have flying cars, but why would we need one. Automated vehicle services will be pervasive, inexpensive, and safe. That is probably a reasonable prediction for 2040.

By 2030, education will be on the verge of transformation with very high quality online classes by the world's best faculty. It will take until 2040 to overcome objections of 14,000 independent schools boards, administrators in 100,000 schools, and two teacher unions. However, quality and economics will eventually prevail.

Entertainment is increasingly dominated by gaming technologies, with the games industry already ten times larger than movies and videos. It seems reasonable to project the latter being absorbed by the former. Movies will be "filmed" in games. Physical movie sets will become increasingly rare.

The sets for these movies will be crafted by AI-based artists. The screenplays will be drafted by Chat GPT capabilities. Humans will oversee all this but there will be far fewer humans involved with the creation of these movies. The remaining humans will have AI-based tools to support these oversight responsibilities.

DOI: 10.4324/9781003491927-10

Nevertheless, there will be an enormous increase in human involvement in these entertainment environments. New themes and capabilities will be crowdsourced. Ideas that are accepted will be well compensated. Ideas that provide sustainable revenue streams will gain residuals for the originators.

Game-Based Transformations

Consider how game-based transformations will affect three domains – education, management, and operations. Active learning will supplant passive learning. Reactive management and operations will become proactive management and operations.

Transforming Education

Table 10.1 summarizes the transformation of education. Experiential, active learning will displace traditional classrooms. The "Sages on the Stage" will be replaced by "Guides on the Side," both human and AI-based. More on this later.

These changes are already happening – see Coursera, edX, and Udacity. High-quality online offerings proved their merits during the pandemic. However, I do not see everyone going to school at home. Students will go to schools that are more like laboratories than lecture halls. They will collaborate with other students both online and in immersive settings such as shown in Figure 8.1.

Transforming Management

This approach to education will prepare students for the world of work. Game-based thinking and practices will transform management as summarized in Table 10.2. Capabilities to explore creative "What if?" questions will expand managers' horizons far beyond incremental variations of the status quo. Game-based predictive capabilities will enable proactive planning, organizing, and control. Participation in addressing management issues and priorities will steadily increase as employees join the games.

The changes depicted in Table 10.2 are already happening. In the future, such transformation will become pervasive. Game-based technologies will enable it.

Table 10.1 Traditional versus Game-Based Education

Teaching	*Traditional (Passive Learning)*	*Game-Based (Active Learning)*
Teaching history	Lectures with slides on people, places, dates	Immersive experiences of events & outcomes
Teaching writing	Drafting and critiquing of personal essays	Formulating & creating interactive stories
Teaching mathematics	Derivations and proofs of equation solutions	Animation of phenomena represented by equations
Teaching science	Lectures with slides on disciplinary phenomena	Simulation of phenomena with supporting data

Table 10.2 Traditional versus Game-Based Management

Function	*Traditional (Reactive Management)*	*Game-Based (Proactive Management)*
Planning	Incremental variations of status quo	Wide range of "What if?" scenarios considered
Organizing	Stewards of status quo prevail – minimal change	Model-based enterprise transformation
Control	Static focus on making the numbers	Predictive model-based control

Table 10.3 Traditional versus Game-Based Operations

Function	*Traditional (Reactive Operations)*	*Game-Based (Proactive Operations)*
Workflow Planning	Incremental variations of status quo	Wide range of "What if?" scenarios considered
Workflow Management	Control charts to manage unacceptable deviations	Predictive model-based management
Supply Chain Management	Siloed management of orders and shipments	Integrated management across supply chain
Finance	Standard discounted cash flow models	Strategic value assessment with options

Transforming Operations

Table 10.3 summarizes the transformation of operations. Execution will be supported by game-based capabilities including predictive models and more dynamic approaches to finance, coupled with integrated views of the enterprise.

These changes have long been underway at more innovative enterprises. They are becoming "table stakes" for all enterprises across public and private ecosystems. Everyone will have operational games that enable maximizing organizational value for customers, employees, and investors.

Artificial Intelligence

How might AI influence the playing of games rather than just the design and development of games as noted in Chapter 9? In many gaming situations, AI will be used to augment human intelligence, rather than replace humans. After all, automation of game playing would defeat the whole purpose of games. What functions are needed to augment intelligence?

Information Management. One function will be information management. This involves information selection (what to present) and scheduling (when to present it). Information modality selection involves choosing among visual, auditory, and tactile channels. Information formatting concerns choosing the best levels of abstraction (concept) and aggregation (detail) for the tasks at hand. AI can be used to make all these choices in real time as the human is pursuing the gaming tasks of interest.

Intent Inferencing. Another function is intent inferencing. Information management can be more helpful if it knows both what humans are doing and what they intend to do. Representing humans' task structure in terms of goals, plans, and scripts can enable making such inferences. Scripts are sequences of actions to which are connected information and control requirements. When the intelligence infers what you intend to do, it then knows what information you need and what controls you want to execute it.

One of the reasons that humans are often included in systems is because they can deal with ambiguity and figure out what to do. Occasionally, what they decide to do has potentially unfortunate consequences. In such cases, "human errors" are reported. Errors in themselves are not the problem. The consequences are the problem. For example, one really bad move can destroy one's prospects in a game.

Error-Tolerant Interfaces. For this reason, another function is an error-tolerant interface. This requires capabilities to identify and classify errors, which are defined as actions that do not make sense (commissions) or the lack of actions (omissions) that seem warranted at the time. Identification and classification lead to remediation. This occurs at three levels: monitoring, feedback, and control. Monitoring involves the collection of more evidence to support the error assessment. Feedback involves making sure the humans realize what they just did. This usually results in humans immediately correcting their errors. Control involves the automation taking over, e.g., applying the brakes, to avoid the imminent consequences.

Adaptive Aiding. The notion of taking control raises the overall issue of whether humans or computers should perform particular tasks. There are many cases where the answer is situation dependent. Thus, this function is termed adaptive aiding. The overall concept is to have mechanisms that enable real time determination of who should be in control. Such mechanisms have been researched extensively, resulting in a framework for design that includes principles of adaptation and principles of interaction.

Intelligent Tutoring. Another function is intelligent tutoring to both train humans and keep them sufficiently in the loop to enable successful human task performance when needed. Training usually addresses two questions: 1) how the system works, and 2) how to work the system. Keeping humans in the loop addresses maintaining competence. Unless tasks can be automated to perfection, humans' competencies need to be maintained. Not surprisingly, this often results in training versus aiding trade-offs, for which guidance has been developed.

Overall Architecture

Figure 10.1 provides an overall architecture for augmenting intelligence. The intelligent interface, summarized above, becomes a component in this broader concept. The overall logic is as follows:

- Humans see displays and controls, and decide and act. Humans need not be concerned with other than these three elements of the architecture. The overall system frames human's roles and tasks, and provides support accordingly.
- The intent inference function infers what task(s) humans intend to do. This function retrieves information and control needs for these task(s). The information management function determines displays and controls appropriate for meeting information and control needs.
- The intelligent tutoring function infers humans' knowledge and skill deficits relative to these task(s). If humans cannot perform the task(s) acceptably, the information management

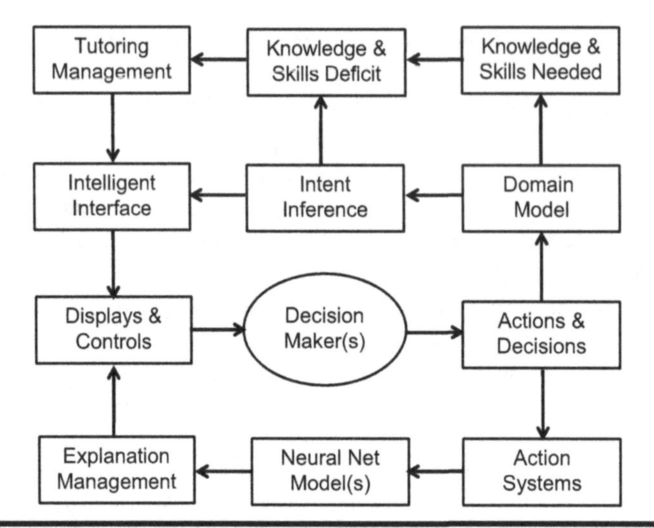

Figure 10.1 Elements of intelligent interface.

function either provides just-in-time training or informs adaptive aiding (see below) of the humans' need for aiding.

- Deep learning neural nets provide recommended actions and decisions. The explanation management function provides explanations of these recommendations to the extent that explanations are requested. This function is elaborated below.
- The adaptive aiding function, within the intelligent interface, determines the human role in execution. This can range from manual to automatic control, with execution typically involving somewhere between these extremes. The error monitoring function, within the intelligent interface, detects, classifies, and remediates anomalies.

Note that these functions influence each other. For example, if adaptive aiding determines that humans should perform task(s), intelligent tutoring assesses the availability of necessary knowledge and skills, and determines the training interventions needed, and information management provides the tutoring experiences to augment knowledge and skills. On the other hand, if adaptive aiding determines that automation should perform task(s), intelligent tutoring assesses humans' abilities to monitor automation, assuming such monitoring is needed.

Explanation Management

Most neural network models cannot explain their (recommended) decisions. This would seem to be a fundamental limitation. However, science has long addressed the need to understand systems that cannot explain their own behaviors. Experimental methods are used to develop statistical models of input-output relationships. Applying these methods to neural network models can yield mathematical models that enable explaining the (recommended) decisions as shown in Figure 10.2.

Given a set of independent variables **X**, a statistical experiment can be designed, e.g., a fractional factorial design, that determines the combinations of values of **X** to be input to the neural net model(s). These models, typically multi-layered, have "learned" from exposure to massive data lakes with labeled instances of true positives, and possibly false positives and false negatives. True negatives are the remaining instances.

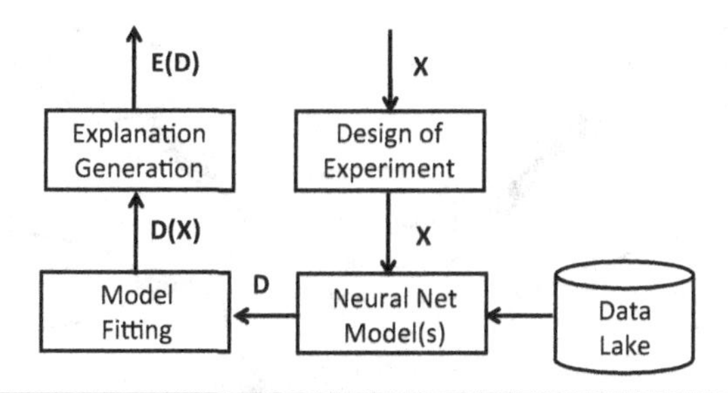

Figure 10.2 Explanation management function.

The neural net models yield decisions, **D**, in response to the designed combinations of **X**. A model **D(X)**, is then fit to these input–output data sets. Explanation generation then yields explanations **E(D)** based on the attributes and weights in the fitted model. The result is a first-order, i.e., non-deep, explanation of the neural net (recommended) decisions.

As noted earlier, the paradigm underlying Figure 10.2 is the standard paradigm of empirical natural science. Thus, it is clear it will work, i.e., yield rule-based explanations, but will it be sufficient to help decision makers understand and accept what the machine learning recommends? We imagine this will depend on the application.

Learning Loops

Figures 10.1 and 10.2 include both explicit and implicit learning loops. The statistical machine-learning loop will be continually refining the relationships in its layers, either by supervised learning or reinforcement learning. This will involve balancing exploration (of uncharted territory) and exploitation (of current knowledge). This may involve human designers and experimenters not included in Figures 10.1 and 10.2. Of particular interest is how machine learning will forget older data and examples that are no longer relevant, e.g., a game strategy that has more recently been shown to be ineffective.

The rule-based learning loops in Figures 10.1 and 10.2 are concerned with inferring rule-based explanations of the recommendations resulting from machine learning (Figure 10.2) and inferring human decision makers' intentions and state of knowledge (Figure 10.1). Further, learning by decision makers is facilitated by the tutoring function in Figure 10.1.

Thus, the AI will be learning about phenomena, cues, decisions, actions, etc. in the overall game environment. The decision makers will learn about what the AI is learning, expressed in more readily understandable rule-based forms. The intelligent support system will be learning about the decision makers' intentions, information needs, etc., as well as influencing what the decision makers learn.

Power of Games

Figure 10.3 summarizes the power of games. Playing, learning, competing, and achievement encourages engagement in active learning, which enhances retention of learning. The competition inherent in games leads to creativity and innovative strategies, plans, and perhaps disruptive game moves.

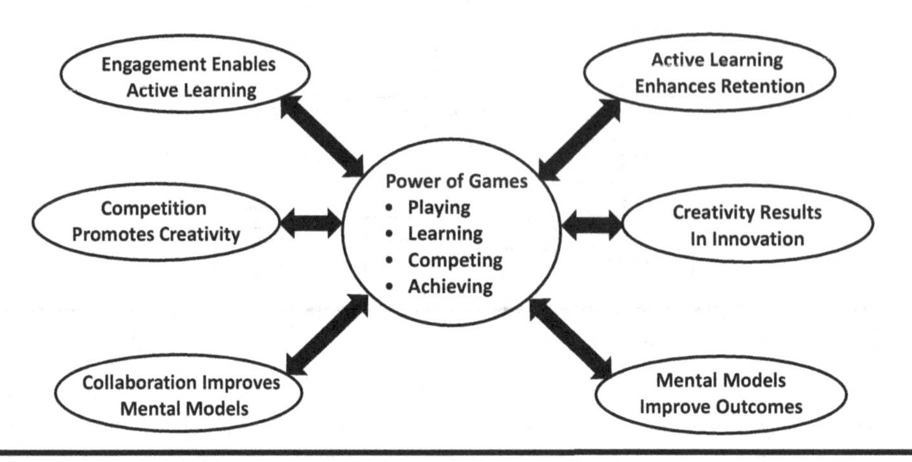

Figure 10.3 The power of games.

This contrasts with the passive filling of trainees' knowledge resources, which they will not remember, nor know how to apply when relevant instances arise. My experiences have included visiting the Parthenon, which was overwhelmingly compelling compared to just reading about it. The power of games includes being there, experiencing this reality, and feeling engaged with how such reality plays out.

Why will all this happen? First and foremost, it is happening. Gaming is a natural human phenomenon, indeed a natural animal phenomenon. The equivalent of natural selection will result in the most appealing and useful gaming practices to prosper. The increasing power and decreasing costs of technology will be a key enabler.

The stewards of the status quo will likely object, arguing that such changes are too risky and expensive. This will likely slow the pace of change, but transformation is inevitable. This reflects society's penchant for "creative destruction," that has long been the foundation of creative prosperity, increasingly for everyone.

Index

Printed in the United States
by Baker & Taylor Publisher Services

Printed in the United States
by Baker & Taylor Publisher Services